Recent Titles in This Series

(Continued in the back of this publication)

MEMOIRS
of the
American Mathematical Society

Number 552

The Major Counting of
Nonintersecting Lattice Paths and
Generating Functions for Tableaux

C. Krattenthaler

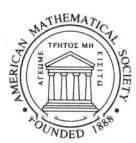

May 1995 • Volume 115 • Number 552 (fourth of 5 numbers) • ISSN 0065-9266

American Mathematical Society
Providence, Rhode Island

1991 *Mathematics Subject Classification.*
Primary 05A15; Secondary 05A10, 05A17, 05A30, 05E10, 33D20.

Library of Congress Cataloging-in-Publication Data

Krattenthaler, C. (Christian), 1958–
    The major counting of nonintersecting lattice paths and generating functions for tableaux /
C. Krattenthaler.
      p. cm. – (Memoirs of the American Mathematical Society, ISSN 0065-9266; no. 552)
    "May 1995, volume 115, number 552, (fourth of 5 numbers)."
    Includes bibliographical references (p. –).
    ISBN 0-8218-2613-1
    1. Generating functions. 2. Lattice paths. 3. Young tableaux. 4. Hypergeometric series.
I. Title. II. Series.
QA3.A57 no. 552
[QA164.8]
510 s–dc20
[511′.62]
                                        95-3815
                                            CIP

Memoirs of the American Mathematical Society

This journal is devoted entirely to research in pure and applied mathematics.

Subscription information. The 1995 subscription begins with Number 541 and consists of six mailings, each containing one or more numbers. Subscription prices for 1995 are $369 list, $295 institutional member. A late charge of 10% of the subscription price will be imposed on orders received from nonmembers after January 1 of the subscription year. Subscribers outside the United States and India must pay a postage surcharge of $25; subscribers in India must pay a postage surcharge of $43. Expedited delivery to destinations in North America $30; elsewhere $92. Each number may be ordered separately; *please specify number* when ordering an individual number. For prices and titles of recently released numbers, see the New Publications sections of the *Notices of the American Mathematical Society.*

    Back number information. For back issues see the *AMS Catalog of Publications.*

    Subscriptions and orders should be addressed to the American Mathematical Society, P. O. Box 5904, Boston, MA 02206-5904. *All orders must be accompanied by payment.* Other correspondence should be addressed to Box 6248, Providence, RI 02940-6248.

Memoirs of the American Mathematical Society is published bimonthly (each volume consisting usually of more than one number) by the American Mathematical Society at 201 Charles Street, Providence, RI 02904-2213. Second-class postage paid at Providence, Rhode Island. Postmaster: Send address changes to Memoirs, American Mathematical Society, P. O. Box 6248, Providence, RI 02940-6248.

Contents

v

Abstract: A theory of counting nonintersecting lattice paths by the major index and generalizations of it is developed. We obtain determinantal expressions for the corresponding generating functions for families of nonintersecting lattice paths with given starting points and given final points, where the starting points lie on a line parallel to $x + y = 0$. In some cases these determinants can be evaluated to result in simple products. As applications we compute the generating function for tableaux with p odd rows, with at most c columns, and with parts between 1 and n. Besides, we compute the generating function for the same kind of tableaux which in addition have only odd parts. We thus also obtain a closed form for the generating function for symmetric plane partitions with at most n rows, with parts between 1 and c, and with p odd entries on the main diagonal. In each case the result is a simple product. By summing with respect to p we provide new proofs of the Bender–Knuth and MacMahon (ex-)Conjectures, which were first proved by Andrews, Gordon, and Macdonald. The link between nonintersecting lattice paths and tableaux is given by variations of the Knuth correspondence.

1991 *Mathematics Subject Classification*: Primary 05A15; Secondary 05A10, 05A17, 05A30, 05E10, 33D20

Keywords: Nonintersecting lattice paths, major index, tableaux, symmetric plane partitions, generating functions, Robinson–Schensted–Knuth correspondence, basic hypergeometric series in $U(n)$ and $Sp(n)$, basic hypergeometric series for A_l and C_l

I. Introduction

Gessel and Viennot's [16, 17] beautiful idea of interpreting tableaux as families of nonintersecting lattice paths is a very useful tool for obtaining determinantal expressions for generating functions for various families of tableaux with a given shape. (Subsequently, Stembridge [41] widened this theory. He showed that generating functions for tableaux where the shape now is allowed to vary can frequently given in form of a Pfaffian.) What is done in this theory, is counting families of nonintersecting lattice paths by weights which depend on the position of the path's edges in the integer lattice $\mathbb{Z}^2$. ($\mathbb{Z}$ stands for the set of integers.) We in turn propose the counting of families of nonintersecting lattice paths by MacMahon's [32] major index and generalizations of it. The original motivation of doing major counting of nonintersecting lattice paths lay in a problem of Choi and Gouyou-Beauchamps. In [7] they found a nice product formula for the number of tableaux with p odd rows, with the parts being bounded by n, and where the lengths of the rows are bounded by $2r$. It was conjectured that there is also a simple product formula for the generating function of this family of tableaux. By a slight modification of Choi and Gouyou-Beauchamps' idea of proof, which is inspired by Desainte-Catherine and Viennot's [10] geometric interpretation of a variation of the celebrated Knuth correspondence [24, 6], in this paper it is shown that the computation of the generating function of the above family of tableaux can be solved by counting nonintersecting lattice paths with given starting and final points, which in addition do not cross the diagonal line $x = y$, by a variation of the major index. So we decided to systematically develop a theory of counting families of nonintersecting lattice paths by major index and generalizations of it, which we call *strange major indices* (cf. (4.0.4)–(4.0.4)).

This major counting theory is developed quite analogously to the usual theory which encounters edge weights. The only difference is that for the major counting we have to assume that the starting points of the lattice paths lie on a line parallel to $x + y = 0$. Once this is done, we also obtain determinants involving q-binomials for the major and strange major generating functions for families of nonintersecting lattice paths with given starting and final points. Also here these determinants in special cases can be evaluated to result in simple products. However, the really significant parts of this paper consider nonintersecting paths which are bounded by $x = y$. This has no counterpart with edge weights. Only the *number* of such families of nonintersecting lattice paths has been previously considered [17, 10, 7]. Again we find determinantal expressions, though a little bit more complicated. And again, in special cases these determinants can be evaluated to result in nice product expressions.

Applications of this theory do not only include the solution of the above tableaux problem. There are other tableaux problems which also can be treated that way. An interesting application of our major counting is to give an alternative proof for the

Received by the editor November 20, 1993.

1

generating function expression for tableaux with rows of even lengths which do not exceed $2r$ and where the parts are bounded by n. (Of course this is the $p = 0$ case of the Choi–Gouyou-Beauchamps problem.) The proofs so far [11, 37, 40] rely on either the theory of symmetric functions or group representation theory. Our proof only uses combinatorial constructions and a simple determinant evaluation. A further by-product of our major counting results is the computation of the generating function for tableaux with p odd rows, with at most c columns, with only odd parts which lie between 1 and $2n - 1$. Equivalently, we thus compute the generating function for symmetric plane partitions with at most n rows, with parts between 1 and c, and with exactly p odd entries on the main diagonal. The $p = 0$ result was also previously given by Désarménien [11, Théorème 1.2, second identity], Proctor [37, Theorem 1 case (CYH)], and Stembridge [40, Corollary 4.3 (b)]. Moreover, having solved the Choi–Gouyou-Beauchamps problem, by summing with respect to p and using the q-analog of Kummer's summation we are able to present a new proof of the Bender–Knuth (ex-)Conjecture [4, p. 50] about the generating function for tableaux with parts between 1 and n, and with at most c columns. Likewise, by summing the generating functions for the above described symmetric plane partitions with respect to p, we provide a new proof of the MacMahon (ex-)Conjecture [32, p. 270] about the generating function for symmetric plane partitions with at most n rows and with parts between 1 and c. Proofs of these conjectures have been given by Andrews [2, 3], Gordon [18, only for the Bender–Knuth conjecture], Macdonald [33, pp. 51–53], and Proctor [36, Propositions 7.2, 7.3]. In addition we prove two theorems (Theorems 9 and 20) about specific generating functions for pairs of tableaux, or, equivalently, for specific sums of squares of Schur functions.

It should be noted that the tableaux and plane partition theorems in this paper could also be derived starting from the summation theorems for Schur functions in Goulden's paper [19] (cf. Remark (2) at the end of section 4). However, it is possible neither to obtain our theorems for the major and strange major counting of noninter-secting lattice paths from Goulden's results nor Goulden's results from our theorems. However, the ideas of the present paper can be suitably modified to even give bijective proofs of Goulden's results. (Goulden used a linear operator for symmetric functions and some tricky manipulations to derive his theorems.) These bijective proofs will appear in a forthcoming paper [29] together with far-reaching generalizations and further applications of the results of Goulden and of the present paper.

We organize the paper in the following way. In the next section most of the definitions are given. The definition of the strange major indices is deferred to section 4, where they appear for the first time. Besides, in subsection 2.1 it is explained how to represent paths as two-rowed arrays, which will be crucial for our combinatorial constructions in subsections 5.1–5.3. In subsection 2.2, as a preparation, we review two sign-reversing involutions which serve to compute the *number* of families of noninter-secting lattice paths. Subsequently, in sections 3 and 4 these involutions will be modified in order to be applicable for the major and strange major counting. In section 3 the counting theory for the major index is developed. Here the results about tableaux with even rows (Theorem 10), about tableaux with odd parts (Theorem 11), respectively symmetric plane partitions with prescribed number of odd entries on the main diagonal (Theorem 12), and the MacMahon Conjecture (Theorem 13) are included. In section 4 the counting theory for the strange major index is developed. Here in Theorem 21 the Choi–Gouyou-Beauchamps problem is solved. The new proof of the Bender–Knuth Conjecture (Theorem 22) is found in this subsection. Since the complete details of the means which we need for the proofs in sections 3 and 4 very often are unpleasant and cumbersome, there is a separate section, namely section 5, which contains all auxiliary results and correspondences. Subsection 5.1 provides the basic results about major and strange major generating functions of single lattice paths. Subsection 5.2 contains two correspondences for paths which cross the line $x = y$. They are "major analogues" for the reflection principle. In subsection 5.3 we give a correspondence for pairs of intersecting lattice paths. It is a "major analogue" for the usual interchanging procedure (cf. [17, 41]). The modified Knuth correspondence together with its geometric interpretation, which was mentioned above, is described in subsection 5.4. The last two subsections are devoted to computational lemmas. A determinant lemma (Lemma 34) of a very general kind is proved in subsection 5.5. Previously, special cases have been successfully used in the computation of plane partition generating functions [25, 26, 28]. Finally, subsection 5.6 provides a multiple summation formula which is of crucial importance in the proofs of Theorems 18 and 19. R. Gustafson and S. Milne pointed out to me that it is a special case of Gustafson's $C_r\ _6\psi_6$ summation (or in the old notation: $Sp(r)\ _6\psi_6$ summation) [21, Theorem 5.1]. The derivation of (5.6.1) from Gustafson's $C_r\ _6\psi_6$ sum can be found in [30]. I have included my original proof of (5.6.1) since it is purely elementary and independent of the deep theory of multiple basic hypergeometric summations and transformations currently under development, mainly by Milne and Gustafson (cf. e.g. [9, 21, 30, 31, 34, 35] and the references cited there). Besides, the ideas in this proof could turn out to be useful for proving other multiple summation formulas. For further comments on the connections with the theory of A_r and C_r summations (formerly: $U(r)$ and $Sp(r)$ summations) see Remark (3) at the end of section 4. For instance, an interesting feature of Theorem 15 is that it implicitly contains a new A_r-type q-Gauß summation formula (identity (4.3.12)).

II. Definitions and Preliminaries

2.1. Definitions and notation. In this paper we always consider lattice paths in the plane consisting of unit horizontal and vertical steps in the positive direction. In the sequel we shall call them *paths* for short. Any path in a natural way corresponds to a multiset permutation consisting of 1's and 2's. Let P be a path from $\mathcal{A} = (A_1, A_2)$ to $\mathcal{E} = (E_1, E_2)$. Later we frequently abbreviate the fact that a path P goes from $\mathcal{A}$ to $\mathcal{E}$ by $P : \mathcal{A} \to \mathcal{E}$. P may be represented by a pair $(\mathcal{A}, \pi)$, where $\mathcal{A}$ is the starting point of P and $\pi = \pi_1 \pi_2 \ldots \pi_{E_1 + E_2 - A_1 - A_2}$, where $\pi_i = 1$ if the i'th step in the path P is a horizontal one and $\pi_i = 2$ if the i'th step in the path P is a vertical one. π is a multiset permutation consisting of $E_1 - A_1$ entries of 1 and $E_2 - A_2$ entries of 2. For example, the path P_0 in Figure 1 is represented by $((1, -1), 221221112122)$. Of course, this representation of paths is unique. Hence, we may identify each path with its representation.

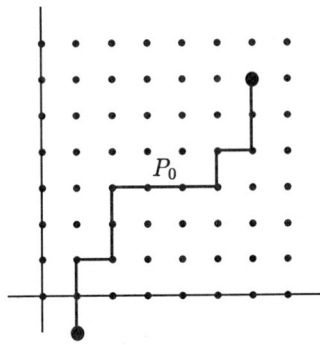

Figure 1

The *major index* (or "greater index") of a multiset permutation $\pi = \pi_1 \pi_2 \ldots \pi_n$, $\pi_i \in \mathbb{N}$ (set of positive integers), is defined by

$$\operatorname{maj} \pi = \sum_{i=1}^{n-1} i \cdot \chi(\pi_i > \pi_{i+1}),$$

where χ is the usual truth function, $\chi(A) = 1$ if A is true, and $\chi(A) = 0$ otherwise.

Given a path $P = (\mathcal{A}, \pi)$, we extend the major index to P by defining $\operatorname{maj} P := \operatorname{maj} \pi$. For our path in Figure 1 we have $\operatorname{maj} P_0 = 2 + 5 + 9 = 16$.

By definition each pair 21 that occurs in a multiset permutation π, and only these, makes a contribution to the major index of π. Given a path $P = (\mathcal{A}, \pi)$, the occurence of 21 in π means that a vertical step is followed by a horizontal one. The point which is the end point of this vertical step (and at the same time the starting point of this horizontal step) will be called a *North-East corner* of the path P. The North-East corners of our path in Figure 1 are $(1, 1)$, $(2, 3)$, and $(5, 4)$. By the above consideration we see that only North-East corners of a path make a contribution to the major index. Besides, the contribution of the North-East corner (a, b) is the number of steps from

4

the starting point of the path to (a, b), or in symbols $a + b - A_1 - A_2$ provided that the starting point is (A_1, A_2). Therefore North-East corners are of great significance when dealing with the major index of paths. In fact, given the starting and the final point of a path, the North-East corners uniquely determine the path. Suppose that P is a path from $\mathcal{A} = (A_1, A_2)$ to $\mathcal{E} = (E_1, E_2)$ and let the North-East corners of P be (a_1, b_1), (a_2, b_2), ..., (a_k, b_k), where we assume that the (a_i, b_i) are ordered from left to right, which is equivalent with $A_1 \leq a_1 < a_2 < \cdots < a_k \leq E_1 - 1$, and $A_2 + 1 \leq b_1 < b_2 < \cdots < b_k \leq E_2$. Then P can be represented by the two-rowed array

$$\begin{matrix} a_1 & a_2 & \ldots & a_k \\ b_1 & b_2 & \ldots & b_k \end{matrix} \, , \tag{2.1.1}$$

or, if we wish to make the bounds which are caused by the starting and the final point transparent,

$$\begin{matrix} A_1 \leq & a_1 \, a_2 \, \ldots \, a_k & \leq E_1 - 1 \\ A_2 + 1 \leq & b_1 \, b_2 \, \ldots \, b_k & \leq E_2 \end{matrix} \, . \tag{2.1.2}$$

For a given starting point and a given final point, by definition the empty array is the representation for the only path that has no North-East corner. For the path in our running example we obtain the array representation

$$\begin{matrix} 1 & 2 & 5 \\ 1 & 3 & 4 \end{matrix} \, ,$$

or with bounds included,

$$\begin{matrix} 1 \leq & 1 \, 2 \, 5 & \leq 5 \\ 0 \leq & 1 \, 3 \, 4 & \leq 6 \end{matrix} \, .$$

The notation (2.1.2) will be used with a three-fold meaning. First, we write down special arrays in that way if in addition we want to point out the row bounds. Secondly, we shall speak of two-rowed arrays *of the type* (2.1.2) by which we will mean arrays of the form (2.1.1) which obey the row bounds in (2.1.2). Thirdly, the notation (2.1.2) will also be used for the *set* of all arrays of the type (2.1.2). It should always be clear from the context which meaning is intended.

In section 5, also two-rowed arrays with its rows being of unequal length will be considered. But these arrays also will have the property that the rows are strictly increasing. So by convention, whenever in this subsection and in subsections 5.1–5.3 we speak of two-rowed arrays we mean two-rowed arrays with strictly increasing rows. Also for these arrays we will use a notation of the kind (2.1.2) together with its three-fold meaning just explained. (We remark that in subsection 5.4 we consider two-rowed arrays of different type.) We shall frequently use the short notation $(a \mid b)$ for two-rowed arrays, where a denotes the sequence (a_i) of elements of the first row, and b denotes the sequence (b_i) of elements of the second row.

Given a sequence $a = a_1, a_2, \ldots, a_k$ we write $\|a\|$ for the sum $\sum_{i=0}^{k} a_i$ of all the elements of the sequence. Also for r-tuples $\eta = (\eta_1, \ldots, \eta_k)$ we shall write $\|\eta\|$ for the sum $\sum_{i=1}^{k} \eta_i$. In order to relate the major index of a path P with starting point

$\mathcal{A} = (A_1, A_2)$ to its array representation $(a \mid b)$, given by (2.1.1), it is readily verified that

$$\operatorname{maj} P = \|a\| + \|b\| - k(A_1 + A_2). \tag{2.1.3}$$

An r-tuple $\lambda = (\lambda_1, \ldots, \lambda_r)$ with $\lambda_1 \geq \lambda_2 \geq \cdots \geq \lambda_r$ is called a *partition*. The components λ_i are called *parts* of the partition. Let λ be a partition. A *tableau* τ of *shape* λ is an array

$$
\begin{array}{llll}
\tau_{11} & \tau_{12} & \cdots\cdots\cdots & \tau_{1\lambda_1} \\
\tau_{21} & \tau_{22} & \cdots\cdots & \tau_{2\lambda_2} \\
\vdots & \cdots\cdots & & \\
\tau_{r1} & \cdots & \tau_{r\lambda_r}
\end{array}
\tag{2.1.4}
$$

of positive integers τ_{ij}, $1 \leq i \leq r$, $1 \leq j \leq \lambda_i$, such that the rows are weakly and the columns are strictly increasing. The entries of τ are called *parts* of the tableau. The sum of all the parts of a tableau τ is called the *norm*, in symbols $n(\tau)$, of the tableau. Given a set T of tableaux, the *norm generating function for T* is defined to be $\sum_{\tau \in T} q^{n(\tau)}$. In the sequel, if we speak of the generating function for some set of tableaux we always mean the norm generating function.

A *plane partition of shape* λ is an array τ of positive integers τ_{ij} of the form (2.1.4) such that the rows and the columns are weakly decreasing. It is called *column-strict* if the columns are strictly decreasing. The notions part, norm, generating function are used for plane partitions in the same sense as with tableaux.

The *Schur function* $s_\lambda(x_1, x_2, \ldots) = s_\lambda(\mathbf{x})$ is a symmetric function (cf. [33, 38, 39]) in the variables $x_1, x_2, \ldots$ and is combinatorially defined by

$$s_\lambda(\mathbf{x}) = \sum_\tau \prod x_{\tau_{ij}} \,,$$

where the sum is over all tableaux τ of shape λ and the product is over all parts τ_{ij} of τ. Equivalently, the sum could also run over all column-strict plane partitions τ of shape λ. The vector $\mathbf{x}$ of variables can be finite or infinite.

Since we deal with generating functions, a short notation for generating functions is in order. We write

$$GF(M; w) := \sum_{x \in M} w(x) \tag{2.1.5}$$

for the generating function of a set M with respect to a weight function w.

The q-notations which will be used are $[\alpha]_q = 1 - q^\alpha$, $[n]_q! = [1]_q [2]_q \cdots [n]_q$, $[0]_q! = 1$,

$$(a; q)_k = \prod_{j=0}^{k-1} (1 - aq^j) \,, \text{ and } (a; q)_0 = 1,$$

$$(a; q)_\infty = \prod_{j=0}^{\infty} (1 - aq^j) \,,$$

so that in particular $[n]_q! = (q;q)_n$, and

$$\begin{bmatrix} n \\ k \end{bmatrix}_q = \begin{cases} \dfrac{[n]_q \cdot [n-1]_q \cdots [n-k+1]_q}{[k]_q!} & k \geq 0 \\ 0 & k < 0 \end{cases}.$$

The base q in $[\alpha]_q$, $[n]_q!$, $(a;q)_k$, $(a;q)_\infty$, and $\begin{bmatrix} n \\ k \end{bmatrix}_q$ will in most cases be omitted. Only if the base is different from q it will be explicitly stated.

The basic hypergeometric series $_{p+1}\phi_p$ is defined by

$$_{p+1}\phi_p \begin{bmatrix} a_1, \ldots, a_{p+1} \\ b_1, \ldots, b_p \end{bmatrix} ; q, z = \sum_{i=0}^{\infty} \frac{(a_1;q)_i (a_2;q)_i \cdots (a_{p+1};q)_i}{(q;q)_i (b_1;q)_i \cdots (b_p;q)_i} z^i.$$

An excellent reference book for basic hypergeometric series is [14]. We will frequently refer to it.

2.2. Two involutions and nonintersecting lattice paths. In order to make the arguments in sections 3 to 5 more transparent, it is perhaps appropriate to recapitulate the usual procedure with nonintersecting lattice paths (cf. [16, 17, 41, section 1]), before going into the major counting theory.

Given two lattice points in the integer lattice $\mathbb{Z}^2$, $\mathcal{A} = (A_1, A_2)$ and $\mathcal{E} = (E_1, E_2)$, we denote the set of all paths from $\mathcal{A}$ to $\mathcal{E}$ by $P(\mathcal{A} \to \mathcal{E})$. For the number of those paths we have

$$|P(\mathcal{A} \to \mathcal{E})| = \binom{E_1 + E_2 - A_1 - A_2}{E_1 - A_1}. \tag{2.2.1}$$

As usual, for a set S by $|S|$ we mean the cardinality of S. Let $\mathcal{A}_i = (A_1^{(i)}, A_2^{(i)})$ and $\mathcal{E}_i = (E_1^{(i)}, E_2^{(i)})$, $i = 1, 2, \ldots, r$, be lattice points. Set $\mathbf{A} := (\mathcal{A}_1, \ldots, \mathcal{A}_r)$ and $\mathbf{E} := (\mathcal{E}_1, \ldots, \mathcal{E}_r)$. The set of all families $(P_1, \ldots, P_r)$ of lattice paths where P_i goes from $\mathcal{A}_i$ to $\mathcal{E}_i$, $i = 1, \ldots, r$, is denoted by $P(\mathbf{A} \to \mathbf{E})$. A family of lattice paths is called *intersecting* if there are two paths in the family which have a point in common, if not the family is called *nonintersecting*. Two *paths* which have a point in common will also be called *intersecting*. The set of families $(P_1, \ldots, P_r)$ of nonintersecting lattice paths where P_i goes from $\mathcal{A}_i$ to $\mathcal{E}_i$ is denoted by $P(\mathbf{A} \to \mathbf{E})^+$, while the set of families $(P_1, \ldots, P_r)$ of intersecting lattice paths is denoted by $P(\mathbf{A} \to \mathbf{E})^-$.

Now let $\mathcal{A}_i, \mathcal{E}_i$ as above and

$$A_1^{(1)} \leq A_1^{(2)} \leq \cdots \leq A_1^{(r)} \quad \text{and} \quad A_2^{(1)} \geq A_2^{(2)} \geq \cdots \geq A_2^{(r)}, \tag{2.2.2a}$$

and

$$E_1^{(1)} \leq E_1^{(2)} \leq \cdots \leq E_1^{(r)} \quad \text{and} \quad E_2^{(1)} \geq E_2^{(2)} \geq \cdots \geq E_2^{(r)}. \tag{2.2.2b}$$

To count the number $|P(\mathbf{A} \to \mathbf{E})^+|$ of all families $(P_1, \ldots, P_r)$ of nonintersecting lattice paths, $P_i : \mathcal{A}_i \to \mathcal{E}_i$, the following procedure is used. Obviously the relation

$$|P(\mathbf{A} \to \mathbf{E})^+| = |P(\mathbf{A} \to \mathbf{E})| - |P(\mathbf{A} \to \mathbf{E})^-| \tag{2.2.3}$$

holds. Let $\mathfrak{S}_r$ be the symmetric group of order r. For a permutation $\sigma \in \mathfrak{S}_r$ define $\mathbf{A}_\sigma$ to be the r-tuple $(\mathcal{A}_{\sigma(1)}, \mathcal{A}_{\sigma(2)}, \ldots, \mathcal{A}_{\sigma(r)})$. In particular, we have $\mathbf{A}_{id} = \mathbf{A}$. The trick is to consider the sets $P(\mathbf{A}_\sigma \to \mathbf{E})^-$, $\sigma \in \mathfrak{S}_r$, i.e. one allows permutations of the starting points. We are going to construct a map Ω which acts on intersecting families. Suppose that the family $\mathfrak{P} = (P_1, \ldots, P_r)$ is an element of $P(\mathbf{A}_\sigma \to \mathbf{E})^-$. Among all meeting points between any two paths of that family consider the right-most meeting points, and among these the highest meeting point. To be precise, if (X, Y) is this meeting point just obtained, then for any other meeting point (x, y) we have

$$x < X, \quad \text{or if } x = X \text{ then } Y \geq y. \tag{2.2.4}$$

Among all pairs of paths which have (X, Y) in common let (P_I, P_J) be that pair for which (I, J) is minimal with respect to the lexicographical order of pairs of integers. To be precise, if (P_i, P_j) is another pair of paths with common point (X, Y) then

$$I < i, \quad \text{or if } I = i \text{ then } J \leq j. \tag{2.2.5}$$

It is a simple matter of fact that by (2.2.4) the point (X, Y) is the last meeting point between P_I and P_J. Denote this last meeting point by $\mathcal{S}$. Next we decompose P_I into $P_I^{(1)}$ and $P_I^{(2)}$, where $P_I^{(1)}$ is the portion of P_I going from $\mathcal{A}_I$ to $\mathcal{S}$, and $P_I^{(2)}$ is the portion of P_I between $\mathcal{S}$ and $\mathcal{E}_I$. Analogously, P_J by $\mathcal{S}$ is decomposed into $P_J^{(1)}$ and $P_J^{(2)}$. Then by concatenation form the new paths $P_I' = P_J^{(1)} P_I^{(2)}$ and $P_J' = P_I^{(1)} P_J^{(2)}$. In other terms, the initial portions of P_I and P_J up to their last meeting point are interchanged, the tails are preserved. Thus we obtain the new family $\Omega(\mathfrak{P}) = (P_1, \ldots, P_I', \ldots, P_J', \ldots, P_r)$. Obviously it is an element of $P(\mathbf{A}_{\sigma(IJ)} \to \mathbf{E})$. $((IJ)$ denotes the transposition which exchanges I and J. $\sigma(IJ)$ is the composition of σ and (IJ).) Obviously this new family is again intersecting, therefore it is an element of $P(\mathbf{A}_{\sigma(IJ)} \to \mathbf{E})^-$. Moreover, since nothing was changed *after* the point $\mathcal{S}$, $\mathcal{S}$ is also the right-most and highest meeting point for the family $\Omega(\mathfrak{P})$, and I and J are also the minimal indices for intersecting pairs of paths of the family $\Omega(\mathfrak{P})$. Therefore when applying the *interchanging procedure* Ω to $\Omega(\mathfrak{P})$, one arrives at $\mathfrak{P}$ again. Consequently Ω is an involution on

$$\bigcup_{\sigma \in \mathfrak{S}_r} P(\mathbf{A}_\sigma \to \mathbf{E})^- , \tag{2.2.6}$$

and moreover a bijection between the subsets

$$\bigcup_{\sigma \text{ even}} P(\mathbf{A}_\sigma \to \mathbf{E})^- \quad \text{and} \quad \bigcup_{\sigma \text{ odd}} P(\mathbf{A}_\sigma \to \mathbf{E})^- .$$

This last fact may be formulated in a different way. On the set

$$\bigcup_{\sigma \in \mathfrak{S}_r} P(\mathbf{A}_\sigma \to \mathbf{E}) , \tag{2.2.7}$$

we introduce the weight function ω by

$$\omega(\mathfrak{P}) = \operatorname{sgn} \sigma \qquad \text{if } \mathfrak{P} \in P(\mathbf{A}_\sigma \to \mathbf{E}).$$

Then Ω is a *sign reversing* involution on the set (2.2.6) with respect to the weight function ω. As usual, Ω being sign-reversing with respect to ω means that $\omega(\Omega(\mathfrak{P})) = -\omega(\mathfrak{P})$. From this we get

$$\sum_{\sigma \in \mathfrak{S}_r} \operatorname{sgn} \sigma \, |P(\mathbf{A}_\sigma \to \mathbf{E})^-| = 0 . \tag{2.2.8}$$

By consecutive use of (2.2.3), (2.2.8), and (2.2.1), and by the observation that by (2.2.2) we have

$$P(\mathbf{A}_\sigma \to \mathbf{E})^- = P(\mathbf{A}_\sigma \to \mathbf{E}) \quad \text{if } \sigma \neq id, \tag{2.2.9}$$

we obtain

$$
\begin{aligned}
|P(\mathbf{A} \to \mathbf{E})^+| &= |P(\mathbf{A} \to \mathbf{E})| - |P(\mathbf{A} \to \mathbf{E})^-| \\
&= |P(\mathbf{A} \to \mathbf{E})| + \sum_{\sigma \in \mathfrak{S}_r, \sigma \neq id} \operatorname{sgn} \sigma \, |P(\mathbf{A}_\sigma \to \mathbf{E})^-| \\
&= \sum_{\sigma \in \mathfrak{S}_r} \operatorname{sgn} \sigma \, |P(\mathbf{A}_\sigma \to \mathbf{E})| \\
&= \sum_{\sigma \in \mathfrak{S}_r} \operatorname{sgn} \sigma \prod_{i=1}^{r} |P(\mathcal{A}_{\sigma(i)} \to \mathcal{E}_i)| \\
&= \sum_{\sigma \in \mathfrak{S}_r} \operatorname{sgn} \sigma \prod_{i=1}^{r} \binom{E_1^{(i)} + E_2^{(i)} - A_1^{(\sigma(i))} - A_2^{(\sigma(i))}}{E_1^{(i)} - A_1^{(\sigma(i))}} \\
&= \det_{1 \leq s, t \leq r} \left(\binom{E_1^{(s)} + E_2^{(s)} - A_1^{(t)} - A_2^{(t)}}{E_1^{(s)} - A_1^{(t)}} \right) . \tag{2.2.10}
\end{aligned}
$$

Since the set of edges trivially does not change under the above described interchanging procedure, this procedure also preserves edge weights (by which we mean weights which depend on the position of the path's edges in $\mathbb{Z}^2$) and therefore can also be used for the counting of intersecting lattice paths with respect to edge weights (see [17, 41, section 1]). But interchanging portions of paths does not at all take care of the major index or strange major index of the paths. So, in subsections 3.1 and 4.1, the interchanging procedure Ω has to be modified in order to work with the major and the strange major index.

In subsections 3.2 and 4.2 we are going to count families of nonintersecting lattice paths with given starting and final points, which in addition do not cross the line $x = y$, by major index and strange major index, respectively. What we are going to use in these sections are modifications of the following procedure which yields the *number* of all those families. Again we are going to construct a sign-reversing involution, which we this time denote by Θ. The idea is to combine the above described interchanging procedure with the *reflection principle* (see e.g. [8, p. 22]). Recall that the reflection principle serves to count all paths from $\mathcal{A} = (A_1, A_2)$ to $\mathcal{E} = (E_1, E_2)$ which do not cross the line $x = y$. Suppose that $A_1 \geq A_2$ and $E_1 \geq E_2$, i.e. that $\mathcal{A}$ and $\mathcal{E}$ lie below or on the line $x = y$. Suppose that we want to count all the paths from $\mathcal{A}$ to $\mathcal{E}$ which *lie below* $x = y$ thereby being allowed to touch $x = y$. If in the sequel we say that "a path lies below the line g" we *always* mean that it is allowed to touch g, the same holding for "above". The reflection principle shows that the paths from $\mathcal{A}$ to $\mathcal{E}$ which *cross* $x = y$ are in one-to-one correspondence with the paths from $(A_2 - 1, A_1 + 1)$ to (E_1, E_2). This is seen by reflecting a crossing path's initial portion up to the *last* point which lies above $x = y$ in the line $x = y - 1$. (This differs from the presentation in [8], where the first point is considered.) The number of paths which lie below $x = y$ (and hence do not cross $x = y$) then is obtained as the difference between the number of *all* paths from (A_1, A_2) to (E_1, E_2) and the number of *all* paths from $(A_2 - 1, A_1 + 1)$ to (E_1, E_2). Let us denote the reflection in the line $x = y - 1$ by $\mathfrak{R}$.

As before, let $\mathcal{A}_i = (A_1^{(i)}, A_2^{(i)})$ and $\mathcal{E}_i = (E_1^{(i)}, E_2^{(i)})$, $i = 1, 2, \ldots, r$, be lattice points which satisfy (2.2.2). Again, set $\mathbf{A} := (\mathcal{A}_1, \ldots, \mathcal{A}_r)$ and $\mathbf{E} := (\mathcal{E}_1, \ldots, \mathcal{E}_r)$. The set of all families $(P_1, \ldots, P_r)$ of lattice paths, where P_i goes from $\mathcal{A}_i$ to $\mathcal{E}_i$, $i = 1, \ldots, r$, this time for convenience is denoted by $Q(\mathbf{A} \to \mathbf{E})$. We want to determine the number of all families $(P_1, \ldots, P_r)$ of nonintersecting lattice paths which do not cross $x = y$, P_i going from $\mathcal{A}_i$ to $\mathcal{E}_i$. The problem only makes sense if all the points $\mathcal{A}_i$ and $\mathcal{E}_i$ lie on one side of the line $x = y$. For the following we assume that the points $\mathcal{A}_i, \mathcal{E}_i$ lie below or on the line $x = y$, i.e.

$$A_1^{(i)} \geq A_2^{(i)} \quad \text{and} \quad E_1^{(i)} \geq E_2^{(i)}, \quad i = 1, 2, \ldots, r. \tag{2.2.11}$$

The set of families $(P_1, \ldots, P_r)$ of nonintersecting lattice paths which lie below $x = y$, P_i going from $\mathcal{A}_i$ to $\mathcal{E}_i$, is denoted by $Q(\mathbf{A} \to \mathbf{E})^+$, while the set of families $(P_1, \ldots, P_r)$ of lattice paths which are either intersecting or contain a path which crosses $x = y$ is denoted by $Q(\mathbf{A} \to \mathbf{E})^-$.

What we want to determine is the number of families of nonintersecting paths lying below $x = y$, i.e. we want to determine the number $|Q(\mathbf{A} \to \mathbf{E})^+|$. Obviously the relation

$$|Q(\mathbf{A} \to \mathbf{E})^+| = |Q(\mathbf{A} \to \mathbf{E})| - |Q(\mathbf{A} \to \mathbf{E})^-| \tag{2.2.12}$$

holds. Just as we above allowed permutations of the starting points, here we allow permutations *and reflections* of the starting points. We extend the reflection in the line $x = y - 1$, $\mathfrak{R}$, to r-tuples of points in the following way. Let η be an r-tuple consisting of 0's and 1's, $\eta \in \{0, 1\}^r$. By $\mathfrak{R}^\eta \mathbf{A}$ we mean the r-tuple $(\mathfrak{R}^{\eta_1} \mathcal{A}_1, \ldots, \mathfrak{R}^{\eta_r} \mathcal{A}_r)$, where of course $\mathfrak{R}^0$ means the identity. Now we consider the sets $Q(\mathfrak{R}^\eta \mathbf{A}_\sigma \to \mathbf{E})^-$. Suppose that the family $\mathfrak{P} = (P_1, \ldots, P_r)$ is an element of $Q(\mathfrak{R}^\eta \mathbf{A}_\sigma \to \mathbf{E})^-$. Among all meeting points between any two paths of that family and all meeting points of any path with the line $x = y - 1$, consider the right-most meeting points, and among these the highest meeting point. Again this is understood in the sense of (2.2.4).

Denote this meeting point by S. Now we distinguish between two cases. If S is a common point of two paths of the family $\mathfrak{P}$, then, as before, among the pairs of paths which have S in common, choose (P_I, P_J) with (I, J) being minimal in the sense of (2.2.5) and apply the above described interchanging procedure, thus obtaining the new family $\Theta(\mathfrak{P}) = (P_1, \ldots, P_I', \ldots, P_J', \ldots, P_r)$. It is an element of $Q(\mathfrak{R}^{\eta^{(IJ)}} \mathbf{A}_{\sigma(IJ)} \to \mathbf{E})^-$, where $\eta^{(IJ)}$ is η with the I'th and J'th coordinate exchanged. To be precise, $\eta_i^{(IJ)} = \eta_i$ for $i \neq I, J$, $\eta_I^{(IJ)} = \eta_J$, and $\eta_J^{(IJ)} = \eta_I$. On the other hand, if S is not a common point then it must lie on $x = y - 1$. Among the paths which go through S choose P_I with I being minimal. Let the image of P_I under application of the reflection principle be denoted by P_I''. Here, "applying the reflection principle to a path P" is understood to mean the reflection of the initial portion of P up to the last meeting point with $x = y - 1$ in the line $x = y - 1$. We then obtain the new family $\Theta(\mathfrak{P}) = (P_1, \ldots, P_I'', \ldots, P_r)$. This family is an element of $Q(\mathfrak{R}^{\eta^{(I)}} \mathbf{A}_\sigma \to \mathbf{E})^-$, where $\eta^{(I)}$ is given by $\eta^{(I)} = (\eta_1, \ldots, 1 - \eta_I, \ldots, \eta_r)$. Just as above, since in both cases nothing was changed *after* the point S, by application of Θ to $\Theta(\mathfrak{P})$, one arrives at $\mathfrak{P}$ again. Consequently Θ is an involution on

$$\bigcup_{\sigma \in \mathfrak{S}_r, \, \eta \in \{0,1\}^r} Q(\mathfrak{R}^\eta \mathbf{A}_\sigma \to \mathbf{E})^- . \tag{2.2.13}$$

Next on the set

$$\bigcup_{\sigma \in \mathfrak{S}_r, \, \eta \in \{0,1\}^r} Q(\mathfrak{R}^\eta \mathbf{A}_\sigma \to \mathbf{E}) . \tag{2.2.14}$$

we introduce the weight ϑ by

$$\vartheta(\mathfrak{P}) := (-1)^{\|\eta\|} \operatorname{sgn} \sigma \qquad \text{if } \mathfrak{P} \in Q(\mathfrak{R}^\eta \mathbf{A}_\sigma \to \mathbf{E})^- .$$

It is easily verified that Θ is sign-reversing with respect to ϑ. From this we infer

$$\sum_{\sigma \in \mathfrak{S}_r, \, \eta \in \{0,1\}^r} (-1)^{\|\eta\|} \operatorname{sgn} \sigma \, |Q(\mathfrak{R}^\eta \mathbf{A}_\sigma \to \mathbf{E})^-| = 0. \tag{2.2.15}$$

By consecutive use of (2.2.12), (2.2.15), and (2.2.1), and by the simple observation that we have

$$Q(\mathfrak{R}^\eta \mathbf{A}_\sigma \to \mathbf{E})^- = Q(\mathfrak{R}^\eta \mathbf{A}_\sigma \to \mathbf{E}) \quad \text{if } (\sigma, \eta) \neq (id, \mathbf{0}) , \tag{2.2.16}$$

we obtain

$$|Q(\mathbf{A} \to \mathbf{E})^+| = |Q(\mathbf{A} \to \mathbf{E})| - |Q(\mathbf{A} \to \mathbf{E})^-|$$

$$= |Q(\mathbf{A} \to \mathbf{E})| + \sum_{\substack{\sigma \in \mathfrak{S}_r,\, \eta \in \{0,1\}^r \\ (\sigma,\eta) \neq (id,\mathbf{0})}} (-1)^{\|\eta\|} \operatorname{sgn} \sigma \, |Q(\mathfrak{R}^\eta \mathbf{A}_\sigma \to \mathbf{E})^-|$$

$$= \sum_{\sigma \in \mathfrak{S}_r,\, \eta \in \{0,1\}^r} (-1)^{\|\eta\|} \operatorname{sgn} \sigma \, |Q(\mathfrak{R}^\eta \mathbf{A}_\sigma \to \mathbf{E})|$$

$$= \sum_{\sigma \in \mathfrak{S}_r} \operatorname{sgn} \sigma \prod_{i=1}^r (|P(\mathcal{A}_{\sigma(i)} \to \mathcal{E}_i)| - |P(\mathfrak{R}\mathcal{A}_{\sigma(i)} \to \mathcal{E}_i)|)$$

$$= \sum_{\sigma \in \mathfrak{S}_r} \operatorname{sgn} \sigma \prod_{i=1}^r \left(\binom{E_1^{(i)} + E_2^{(i)} - A_1^{(\sigma(i))} - A_2^{(\sigma(i))}}{E_1^{(i)} - A_1^{(\sigma(i))}} \right.$$
$$\left. - \binom{E_1^{(i)} + E_2^{(i)} - A_1^{(\sigma(i))} - A_2^{(\sigma(i))}}{E_2^{(i)} - A_1^{(\sigma(i))} - 1} \right)$$

$$= \det_{1 \leq s,t \leq r} \left(\binom{E_1^{(s)} + E_2^{(s)} - A_1^{(t)} - A_2^{(t)}}{E_1^{(s)} - A_1^{(t)}} - \binom{E_1^{(s)} + E_2^{(s)} - A_1^{(t)} - A_2^{(t)}}{E_2^{(s)} - A_1^{(t)} - 1} \right).$$

$$(2.2.17)$$

III. Counting by Major Index

In this section we want to count families $\mathfrak{P} = (P_1, \ldots, P_r)$ of lattice paths by the major index. To be precise, we want to count families $\mathfrak{P}$ by $\mathrm{maj}\,\mathfrak{P}$ where the major index is extended to families of paths by

$$\mathrm{maj}\,\mathfrak{P} = \sum_{i=1}^{r} \mathrm{maj}\,P_r. \tag{3.0.1}$$

3.1. Without restrictions. We precede our first theorem by an auxiliary lemma.

Lemma 1. *Let $\mathcal{A}_i = (A_1^{(i)}, A_2^{(i)})$ and $\mathcal{E}_i = (E_1^{(i)}, E_2^{(i)})$, $i = 1, 2, \ldots, r$, be lattice points in the integer lattice $\mathbb{Z}^2$ which satisfy (2.2.2). Given a permutation $\sigma \in \mathfrak{S}_r$, suppose that $\mathfrak{P} = (P_1, \ldots, P_r)$ is an intersecting family of lattice paths that is an element of $P(\mathbf{A}_\sigma \to \mathbf{E})$. Then there exists an index I such that P_I and P_{I+1} intersect.*

PROOF. Let P_J and P_K, $J < K$, be two intersecting paths in the family $\mathfrak{P}$ such that the difference $K - J$ is minimal. If $K - J = 1$ we are done. Now suppose that $K - J > 1$. Everything becomes clear if Figure 2 is considered.

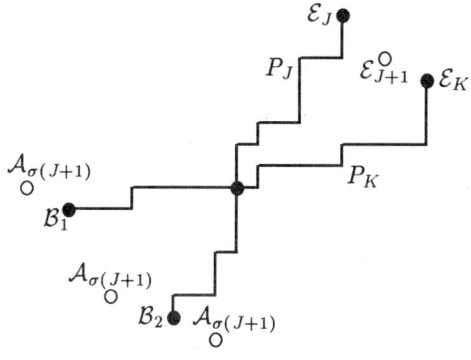

Figure 2

The pair of points $(\mathcal{B}_1, \mathcal{B}_2)$ is meant to be identical either with $(\mathcal{A}_{\sigma(J)}, \mathcal{A}_{\sigma(K)})$ or with its reversion $(\mathcal{A}_{\sigma(K)}, \mathcal{A}_{\sigma(J)})$, depending on whether $\sigma(J) < \sigma(K)$ or not. Because of (2.2.2b) the final point $\mathcal{E}_K$ of P_K must be located in the South-East of $\mathcal{E}_J$, as well as because of (2.2.2a) the starting point $\mathcal{A}_{\sigma(K)}$ of P_K must be located in the South-East or North-West of $\mathcal{A}_{\sigma(J)}$. Moreover, because of (2.2.2a) the starting point $\mathcal{A}_{\sigma(J+1)}$ of P_{J+1} must be located either in the North-West of $\mathcal{B}_1$, or in the South-East of $\mathcal{B}_1$ and at the same time in the North-West of B_2, or in the South-East of $\mathcal{B}_2$. The final point $\mathcal{E}_{J+1}$ of P_{J+1} must be located in the South-East of $\mathcal{E}_J$ and in the North-West of $\mathcal{E}_K$. In Figure 2 typical positions of $\mathcal{A}_{\sigma(J+1)}$ and $\mathcal{E}_{J+1}$ are marked by circles. Obviously the path P_{J+1} has to intersect either P_J or P_K in order to connect $\mathcal{A}_{\sigma(J+1)}$ and $\mathcal{E}_{J+1}$. This is a contradiction to the minimality of $K - J$. $\square$

We call paths P_I, P_{I+1} with consecutive indices *neighbouring paths*. Now we are in the position to state our first result.

Theorem 2. Let $\mathcal{A}_i = (A^{(i)} + D, -A^{(i)})$ and $\mathcal{E}_i = (E_1^{(i)}, E_2^{(i)})$, $i = 1, 2, \ldots, r$, be lattice points in the integer lattice $\mathbb{Z}^2$ with

$$A^{(1)} < A^{(2)} < \cdots < A^{(r)} , \tag{3.1.1a}$$

and

$$E_1^{(1)} < E_1^{(2)} < \cdots < E_1^{(r)} \quad \text{and} \quad E_2^{(1)} \geq E_2^{(2)} \geq \cdots \geq E_2^{(r)} . \tag{3.1.1b}$$

The generating function $\sum q^{\mathrm{maj}\,\mathfrak{P}}$, where the sum is over all nonintersecting families $\mathfrak{P} = (P_1, \ldots, P_r)$ of lattice paths, $P_i : \mathcal{A}_i \to \mathcal{E}_i$, $i = 1, 2, \ldots, r$, and where $\mathrm{maj}\,\mathfrak{P}$ is given by (3.0.1), is equal to the expression

$$\det_{1 \leq s, t \leq r} \left(q^{s(A^{(s)} - A^{(t)})} \begin{bmatrix} E_1^{(s)} + E_2^{(s)} - D \\ E_1^{(s)} - A^{(t)} - D \end{bmatrix} \right). \tag{3.1.2}$$

REMARK. In subsection 5.3 it is explained why for major counting we have to require that the starting points of the paths have to lie on a line parallel to $x + y = 0$.

PROOF OF THE THEOREM. We adopt the notations of subsection 2.2. First we define a weight function ω_1 on the set (2.2.7) by

$$\omega_1(\mathfrak{P}) = \operatorname{sgn}\sigma \, q^{\sum_{i=1}^{r} i(A^{(i)} - A^{(\sigma(i))})} q^{\mathrm{maj}\,\mathfrak{P}} \qquad \text{if } \mathfrak{P} \in P(\mathbf{A}_\sigma \to \mathbf{E}). \tag{3.1.3}$$

In terms of ω_1 the generating function in question is $GF(P(\mathbf{A} \to \mathbf{E})^+; \omega_1)$, where the symbol $GF(M; w)$ is explained in (2.1.5). Obviously, the equation

$$GF(P(\mathbf{A} \to \mathbf{E})^+; \omega_1) = GF(P(\mathbf{A} \to \mathbf{E}); \omega_1) - GF(P(\mathbf{A} \to \mathbf{E})^-; \omega_1) \tag{3.1.4}$$

holds, which is the analogue for (2.2.3).

Similar to subsection 2.2, we are going to construct an involution Ω_1 on the set (2.2.6) which is sign-reversing with respect to ω_1. Let $\mathfrak{P} = (P_1, \ldots, P_r)$ be an element of $P(\mathbf{A}_\sigma \to \mathbf{E})^-$. Consider all meeting points between *neighbouring* paths. That there is at least one is guaranteed by Lemma 1. (This is more restrictive than in subsection 2.2 where we considered the meeting points between *any* two paths.) Choose the right-most, and among them, the highest meeting point and denote it by $\mathcal{S}$. Among the pairs P_I, P_{I+1} which meet in $\mathcal{S}$ choose that one with I being minimal. It should be noted that $\mathcal{S}$ is the last meeting point between P_I and P_{I+1}. Then we define

$$\Omega_1(\mathfrak{P}) := (P_1, \ldots, P_I', P_{I+1}', \ldots, P_r) \tag{3.1.5a}$$

where the paths P_I', P_{I+1}' are given by

$$(P_I', P_{I+1}') = \begin{cases} \Psi((P_I, P_{I+1})) & \text{if } \sigma(I) < \sigma(I+1) \\ \Psi^{(-1)}((P_I, P_{I+1})) & \text{if } \sigma(I) > \sigma(I+1) \end{cases}. \tag{3.1.5b}$$

The map Ψ is defined in Proposition 27. It is readily observed that $\Omega_1(\mathfrak{P})$ is an element of $P(\mathbf{A}_{\sigma(I, I+1)} \to \mathbf{E})^-$. Besides, since Ψ is constructed just in the manner that it does not change anything after this last meeting point $\mathcal{S}$, application of Ω_1 to $\Omega_1(\mathfrak{P})$ gives $\mathfrak{P}$ again. Hence, Ω_1 is an involution on the set (2.2.7).

Next we show that Ω_1 is sign-reversing with respect to ω_1. Namely, by the definitions (3.1.3), (3.1.5), and using (5.3.4), we get

$$
\begin{aligned}
\omega_1(\Omega_1(\mathfrak{P})) &= \operatorname{sgn}\left(\sigma(I, I+1)\right) q^{\sum_{i=1}^{r} i(A^{(i)} - A^{((\sigma(I,I+1))(i)))}} q^{\operatorname{maj} \Omega_1(\mathfrak{P})} \\
&= -\operatorname{sgn}\sigma \, q^{\sum_{i=1}^{r} i(A^{(i)} - A^{(\sigma(i))}) + I(A^{(\sigma(I))} - A^{(\sigma(I+1))}) + (I+1)(A^{(\sigma(I+1))} - A^{(\sigma(I))})} \\
&\quad \cdot q^{\operatorname{maj} \mathfrak{P} + A^{(\sigma(I))} - A^{(\sigma(I+1))}} \\
&= -\omega_1(\mathfrak{P}).
\end{aligned}
\tag{3.1.6}
$$

This implies the identity

$$
\sum_{\sigma \in \mathfrak{S}_r} GF(P(\mathbf{A}_\sigma \to \mathbf{E})^-; \omega_1) = 0 .
\tag{3.1.7}
$$

Obviously, this is the analogue for (2.2.8).

The rest is routine after having read the first part of subsection 2.2. We only have to mimic the computation (2.2.10). Indeed, by consecutive use of (3.1.5), (3.1.7), (2.2.9), and (5.1.13) (the analogue for (2.2.1)), we obtain

$$
\begin{aligned}
GF(P(\mathbf{A} \to \mathbf{E})^+; \omega_1) &= GF(P(\mathbf{A} \to \mathbf{E}); \omega_1) - GF(P(\mathbf{A} \to \mathbf{E})^-; \omega_1) \\
&= GF(P(\mathbf{A} \to \mathbf{E}); \omega_1) + \sum_{\sigma \in \mathfrak{S}_r, \, \sigma \neq id} GF(P(\mathbf{A}_\sigma \to \mathbf{E})^-; \omega_1) \\
&= \sum_{\sigma \in \mathfrak{S}_r} GF(P(\mathbf{A}_\sigma \to \mathbf{E}); \omega_1) \\
&= \sum_{\sigma \in \mathfrak{S}_r} \operatorname{sgn}\sigma \prod_{i=1}^{r} q^{i(A^{(i)} - A^{(\sigma(i))})} GF(P(\mathcal{A}_{\sigma(i)} \to \mathcal{E}_i); \operatorname{maj}) \\
&= \sum_{\sigma \in \mathfrak{S}_r} \operatorname{sgn}\sigma \prod_{i=1}^{r} q^{i(A^{(i)} - A^{(\sigma(i))})} \begin{bmatrix} E_1^{(i)} + E_2^{(i)} - D \\ E_1^{(i)} - A^{(\sigma(i))} - D \end{bmatrix} \\
&= \det_{1 \leq s, t \leq r} \left(q^{s(A^{(s)} - A^{(t)})} \begin{bmatrix} E_1^{(s)} + E_2^{(s)} - D \\ E_1^{(s)} - A^{(t)} - D \end{bmatrix} \right),
\end{aligned}
\tag{3.1.8}
$$

which settles (3.1.2). $\quad\square$

It is very unlikely that the determinant in (3.1.2) can be simplified significantly in general. But there is a special case in which Lemma 35 can be applied to obtain a closed form for the determinant.

Theorem 3. Let $\mathcal{A}_i = (A^{(i)} + D, -A^{(i)})$ and $\mathcal{E}_i = (E_1 + i, E_2 - i)$, $i = 1, 2, \ldots, r$, be lattice points in the integer lattice $\mathbb{Z}^2$ such that (3.1.1a) holds. The generating *function $\sum q^{\mathrm{maj}(\mathfrak{P})}$ where the sum is over all nonintersecting families $\mathfrak{P} = (P_1, \ldots, P_r)$ of lattice paths, $P_i : \mathcal{A}_i \to \mathcal{E}_i$, $i = 1, 2, \ldots, r$, is equal to the expression*

$$\prod_{i=1}^{r} \frac{[E_1 + E_2 + i - D - 1]!}{[E_2 + A^{(i)} - 1]! \, [E_1 + r - D - A^{(i)}]!} \prod_{1 \leq i < j \leq r} [A^{(j)} - A^{(i)}] . \qquad (3.1.9)$$

PROOF. Setting $E_1^{(i)} = E_1 + i$ and $E_2^{(i)} = E_2 - i$ in Theorem 2, we obtain for the desired generating function the expression

$$\det \left(q^{s(A^{(s)} - A^{(t)})} \begin{bmatrix} E_1 + E_2 - D \\ E_1 + s - A^{(t)} - D \end{bmatrix} \right) .$$

Taking some factors out of the determinant leads to

$$\det \left(q^{s(A^{(s)} - A^{(t)})} \begin{bmatrix} E_1 + E_2 - D \\ E_1 + s - A^{(t)} - D \end{bmatrix} \right)$$

$$= q^{\sum_{i=1}^{r} i A^{(i)}} \prod_{i=1}^{r} \frac{[E_1 + E_2 - D]!}{[E_1 + r - A^{(i)} - D]! \, [E_2 + A^{(i)} - 1]!}$$

$$\times \det(q^{-s A^{(t)}} [E_1 + r - D - A^{(t)}] \cdots [E_1 + s + 1 - D - A^{(t)}]$$

$$\cdot [E_2 - s + 1 + A^{(t)}] \cdots [E_2 - 1 + A^{(t)}])$$

$$= (-1)^{\binom{s}{2}} q^{\sum_{i=1}^{r} \left((i-1) A^{(i)} + (E_1 + r - D) + \cdots + (E_1 + i + 1 - D) \right)}$$

$$\times \prod_{i=1}^{r} \frac{[E_1 + E_2 - D]!}{[E_1 + r - D - A^{(i)}]! \, [E_2 + A^{(i)} - 1]!}$$

$$\times \det \left((q^{-E_1 - r - D} - q^{-A^{(t)}}) \cdots (q^{-E_1 - s - 1 - D} - q^{-A^{(t)}}) \right.$$

$$\left. \cdot (q^{E_2 - s + 1} - q^{-A^{(t)}}) \cdots (q^{E_2 - 1} - q^{-A^{(t)}}) \right).$$

Now Lemma 35 with $X_t = -q^{-A^{(t)}}$, $A_s = q^{-E_1 - s - D}$, $p_{j-1}(X) = \prod_{i=2}^{j}(q^{E_2 - i + 1} + X)$ can be applied. After a little bit of manipulation one arrives at (3.1.9). $\qquad\square$

3.2. With a diagonal boundary. In this subsection we count nonintersecting families of paths with fixed starting and final points which do not cross the line $x = y$ with respect to the major index. We assume that the starting points lie on the line $x + y = D$ (see the Remark in the previous subsection). Shifts in the direction (v, v) show that in fact there are two classes of locations of the starting points, one if D is even the other if D is odd. But it is more convenient to state our results for generic D.

First we turn to the case that the paths lie below $x = y$. We use the notations of subsection 2.2. In particular we assume that there is an r-tuple $\mathbf{A} = (\mathcal{A}_1, \ldots, \mathcal{A}_r)$ of starting points and an r-tuple $\mathbf{E} = (\mathcal{E}_1, \ldots, \mathcal{E}_r)$ of final points satisfying (2.2.2) and (2.2.11). Let us introduce the following phrases. Given a family $\mathfrak{P} = (P_1, \ldots, P_r)$

that is an element of $Q(\mathfrak{R}^\eta \mathbf{A}_\sigma \to \mathbf{E})$ we say that the path P_i is of *ordinary type* if $\eta_i = 0$ or, equivalently, if P_i starts at $\mathcal{A}_{\sigma(i)}$. We say that P_i is of *reflected type* if $\eta_i = 1$ or, equivalently, if P_i starts at $\mathfrak{R}\mathcal{A}_{\sigma(i)}$.

Again we precede the general theorem by an auxilary lemma. There are two statements in the lemma. The first will be needed for the proof of the next theorem, the stronger and more technical second statement will be used in the proof of Theorem 16.

Lemma 4. *Let* $\mathcal{A}_i = (A_1^{(i)}, A_2^{(i)})$ *and* $\mathcal{E}_i = (E_1^{(i)}, E_2^{(i)})$, $i = 1, 2, \ldots, r$, *be lattice points in the integer lattice* $\mathbb{Z}^2$ *which satisfy (2.2.2) and (2.2.11). Given a permutation* $\sigma \in \mathfrak{S}_r$ *and* $\eta \in \{0,1\}^r$, *suppose that* $\mathfrak{P} = (P_1, \ldots, P_r)$ *is a family of lattice paths that is an element of* $Q(\mathfrak{R}^\eta \mathbf{A}_\sigma \to \mathbf{E})^-$. *Then either* P_1 *meets* $x = y - 1$, *or else there exists an index* I *such that* P_I *and* P_{I+1} *intersect. Even stronger, either* P_1 *meets* $x = y - 1$, *or else there exists an index* I *such that* P_I *and* P_{I+1} *intersect, where* P_I *is of ordinary type and the paths* $P_1, \ldots, P_I$ *are pairwise nonintersecting and lie below* $x = y$.

PROOF. If P_1 crosses $x = y$, and hence meets $x = y - 1$, then we are done. Therefore let us assume that P_1 lies below $x = y$. Because of (2.2.11) P_1 must be of ordinary type. Suppose that P_2 intersects P_1. Then P_1, P_2 intersect and P_1 is of ordinary type and lies below $x = y$. Hence the theorem would be proved. On the other hand, if P_2 does not intersect P_1 then because of (2.2.2b) it must entirely lie below P_1. Therefore, because of (2.2.2a), P_2 must be of ordinary type. Moreover, lying below the path P_1 which lies below $x = y$, it cannot reach $x = y$ and hence also lies below $x = y$. Now we continue considering P_3, P_4, etc. This procedure must terminate since $\mathfrak{P}$ is intersecting or contains a path which crosses $x = y$. Thus we obtain an intersecting pair P_I, P_{I+1}, where P_I is of ordinary type and where $P_1, \ldots, P_I$ are pairwise nonintersecting and lie below $x = y$. $\square$

Theorem 5. *Let* $\mathcal{A}_i = (A^{(i)} + D, -A^{(i)})$ *and* $\mathcal{E}_i = (E_1^{(i)}, E_2^{(i)})$, $i = 1, 2, \ldots, r$, *be lattice points in the integer lattice* $\mathbb{Z}^2$ *such that (3.1.1) and*

$$2A^{(i)} + D \geq 0 \quad \text{and} \quad E_1^{(i)} \geq E_2^{(i)}, \quad i = 1, 2, \ldots, r,$$

hold. The generating function $\sum q^{\mathrm{maj}(\mathfrak{P})}$, *where the sum is over all nonintersecting families* $\mathfrak{P} = (P_1, \ldots, P_r)$ *of lattice paths which lie below the line* $x = y$, $P_i : \mathcal{A}_i \to \mathcal{E}_i$, $i = 1, 2, \ldots, r$, *and where* $\mathrm{maj}\,\mathfrak{P}$ *is given by (3.0.1), is equal to the expression*

$$\det_{1 \leq s, t \leq r} \left(q^{s(A^{(s)} - A^{(t)})} \left(\begin{bmatrix} E_1^{(s)} + E_2^{(s)} - D \\ E_1^{(s)} - A^{(t)} - D \end{bmatrix} \right. \right.$$
$$\left. \left. - q^{s(2A^{(t)} + D + 1)} \begin{bmatrix} E_1^{(s)} + E_2^{(s)} - D \\ E_2^{(s)} - A^{(t)} - D - 1 \end{bmatrix} \right) \right). \quad (3.2.1)$$

PROOF. Again we use the notations of subsection 2.2. On the set (2.2.14) we introduce the weight function ϑ_1 by

$$\vartheta_1(\mathfrak{P}) = (-1)^{\|\eta\|} \operatorname{sgn} \sigma \, q^{\sum_{i=1}^r \left(i(A^{(i)} - A^{(\sigma(i))}) + i\eta_i(2A^{(\sigma(i))} + D + 1) \right)} q^{\mathrm{maj}\,\mathfrak{P}}$$

$$\text{if } \mathfrak{P} \in Q(\mathfrak{R}^\eta \mathbf{A}_\sigma \to \mathbf{E}). \quad (3.2.2)$$

The generating function in question is $GF(Q(\mathbf{A} \to \mathbf{E})^+; \vartheta_1)$. (The symbol $GF(M; w)$ is explained in (2.1.5).) Obviously, the equation

$$GF(Q(\mathbf{A} \to \mathbf{E})^+; \vartheta_1) = GF(Q(\mathbf{A} \to \mathbf{E}); \vartheta_1) - GF(Q(\mathbf{A} \to \mathbf{E})^-; \vartheta_1) \qquad (3.2.3)$$

holds, which is the analogue of (2.2.12). Similar to subsection 2.2, we are going to construct an involution Θ_1 on the set (2.2.13) which is sign-reversing with respect to ϑ_1. Let $\mathfrak{P} = (P_1, \ldots, P_r)$ be an element of $Q(\mathfrak{R}^\eta \mathbf{A}_\sigma \to \mathbf{E})^-$. Consider all meeting points between neighbouring paths and all meeting points of the path P_1 with the line $x = y - 1$. That there is at least one is guaranteed by Lemma 4. Choose the right-most, and, among them, the highest meeting point, and denote it by $\mathcal{S}$.

Now there are two cases. If $\mathcal{S}$ is a meeting point between two neighbouring paths, then among the pairs P_I, P_{I+1} which meet in $\mathcal{S}$ choose that one with I being minimal. As before, it should be noted that $\mathcal{S}$ is the last meeting point between P_I and P_{I+1}. Then we define

$$\Theta_1(\mathfrak{P}) := (P_1, \ldots, P_I', P_{I+1}', \ldots, P_r) , \qquad (3.2.4a)$$

where the paths P_I', P_{I+1}' are given by

$$(P_I', P_{I+1}') = \begin{cases} \Psi((P_I, P_{I+1})) & \text{if } \sigma(I) < \sigma(I+1) \\ \Psi^{-1}((P_I, P_{I+1})) & \text{if } \sigma(I) > \sigma(I+1) \end{cases} . \qquad (3.2.4b)$$

The map Ψ is defined in Proposition 27. It is readily observed that with (3.2.4a,b) the family $\Theta_1(\mathfrak{P})$ is an element of $Q(\mathfrak{R}^{\eta^{(I,I+1)}} \mathbf{A}_{\sigma(I,I+1)} \to \mathbf{E})^-$, where $\eta^{(I,I+1)} = (\eta_1, \ldots, \eta_{I+1}, \eta_I, \ldots, \eta_r)$.

In case that $\mathcal{S}$ is no meeting point between any two neighbouring paths, $\mathcal{S}$ must be the last meeting point of the path P_1 with the line $x = y - 1$. Then we define

$$\Theta_1(\mathfrak{P}) := (P_1'', P_2, \ldots, P_r), \qquad (3.2.4c)$$

where the path P_1'' is given by

$$P_1'' = \begin{cases} \Phi^{(1)}(P_1) & \text{if } \eta_1 = 0 \\ \left(\Phi^{(1)}\right)^{-1}(P_1) & \text{if } \eta_1 = 1. \end{cases} \qquad (3.2.4d)$$

The map $\Phi^{(1)}$ is defined in Corollary 25. Obviously with (3.2.4c,d) the family $\Theta_1(\mathfrak{P})$ is an element of $Q(\mathfrak{R}^{\eta^{(1)}} \mathbf{A}_\sigma \to \mathbf{E})^-$, where $\eta^{(1)} = (1 - \eta_1, \eta_2, \ldots, \eta_r)$.

Since both Ψ and $\Phi^{(1)}$ are constructed in a manner that they do not change anything after this last meeting point $\mathcal{S}$, application of Θ_1 to $\Theta_1(\mathfrak{P})$ gives $\mathfrak{P}$ again. Hence, Θ_1 is an involution on the set (2.2.13).

We want to establish that Θ_1 is sign-reversing with respect to ϑ_1. Recalling that the starting point of P_i is $\mathfrak{R}^{\eta_i} \mathcal{A}_{\sigma(i)}$, we observe that it can be written in the form

$$(A^{(\sigma(i))} + D, -A^{(\sigma(i))}) + \eta_i(-2A^{(\sigma(i))} - D - 1, 2A^{(\sigma(i))} + D + 1). \qquad (3.2.5)$$

Now, if $\Theta_1(\mathfrak{P})$ is obtained using (3.2.4a,b), by the definition (3.2.2), and using (5.3.4) and (3.2.5), we get

$$
\begin{aligned}
\vartheta_1(\Theta_1(\mathfrak{P})) &= (-1)^{\|\eta^{(I,I+1)}\|} \operatorname{sgn}\left(\sigma(I,I+1)\right) \\
&\quad \cdot q^{\sum_{i=1}^{r}\left(i(A^{(i)}-A^{((\sigma(I,I+1))(i))})+i\eta_i^{(I,I+1)}(2A^{((\sigma(I,I+1))(i))}+D+1)\right)} q^{\operatorname{maj}\Theta_1(\mathfrak{P})} \\
&= -(-1)^{\|\eta\|}\operatorname{sgn}\sigma\, q^{\sum_{i=1}^{r} i(A^{(i)}-A^{(\sigma(i))})+I(A^{(\sigma(I))}-A^{(\sigma(I+1))})+(I+1)(A^{(\sigma(I+1))}-A^{(\sigma(I))})} \\
&\quad \cdot q^{\sum_{i=1}^{r} i\eta_i(2A^{(\sigma(i))}+D+1)-\eta_{I+1}(2A^{(\sigma(I+1))}+D+1)+\eta_I(2A^{(\sigma(I))}+D+1)} \\
&\quad \cdot q^{\operatorname{maj}\mathfrak{P}+\left(A^{(\sigma(I))}-\eta_I(2A^{(\sigma(I))}+D+1)\right)-\left(A^{(\sigma(I+1))}-\eta_{I+1}(2A^{(\sigma(I+1))}+D+1)\right)} \\
&= -\vartheta_1(\mathfrak{P}).
\end{aligned}
\tag{3.2.6}
$$

On the other hand, if $\Theta_1(\mathfrak{P})$ is obtained using (3.2.4c,d), by the definition (3.2.2), and using (3.2.5) and (5.2.20), we get

$$
\begin{aligned}
\vartheta_1(\Theta_1(\mathfrak{P})) &= (-1)^{\|\eta^{(1)}\|}\operatorname{sgn}\sigma\, q^{\sum_{i=1}^{r}\left(i(A^{(i)}-A^{(\sigma(i))})+i\eta_i^{(1)}(2A^{(\sigma(i))}+D+1)\right)} q^{\operatorname{maj}\Theta_1(\mathfrak{P})} \\
&= -(-1)^{\|\eta\|}\operatorname{sgn}\sigma\, q^{\sum_{i=1}^{r} i(A^{(i)}-A^{(\sigma(i))})} q^{\sum_{i=1}^{r} i\eta_i(2A^{(\sigma(i))}+D+1)+(1-2\eta_1)(2A^{(\sigma(1))}+D+1)} \\
&\quad \cdot q^{\operatorname{maj}\mathfrak{P}-\left(A^{(\sigma(1))}+D-\eta_1(2A^{\sigma(1)}+D+1)\right)+\left(-A^{(\sigma(1))}+\eta_1(2A^{\sigma(1)}+D+1)\right)-1} \\
&= -\vartheta_1(\mathfrak{P}).
\end{aligned}
\tag{3.2.7}
$$

This implies the identity

$$
\sum_{\sigma\in\mathfrak{S}_r,\,\eta\in\{0,1\}^r} GF(Q(\mathfrak{R}^\eta \mathbf{A}_\sigma \to \mathbf{E})^-;\vartheta_1) = 0 .
\tag{3.2.8}
$$

Obviously, this is the analogue for (2.2.15).

The rest is routine in view of the second part of subsection 2.2. This time we only have to mimic the computation (2.2.17). Indeed, by consecutive use of (3.2.3), (3.2.8),

(2.2.16), and (5.1.13), we obtain

$$GF(Q(\mathbf{A} \to \mathbf{E})^+; \vartheta_1) = GF(Q(\mathbf{A} \to \mathbf{E}); \vartheta_1) - GF(Q(\mathbf{A} \to \mathbf{E})^-; \vartheta_1)$$

$$= GF(Q(\mathbf{A} \to \mathbf{E}); \vartheta_1) + \sum_{\substack{\sigma \in \mathfrak{S}_r, \, \eta \in \{0,1\}^r \\ (\sigma, \eta) \neq (id, \mathbf{0})}} GF(Q(\mathfrak{R}^\eta \mathbf{A}_\sigma \to \mathbf{E})^-; \vartheta_1)$$

$$= \sum_{\sigma \in \mathfrak{S}_r, \, \eta \in \{0,1\}^r} GF(Q(\mathfrak{R}^\eta \mathbf{A}_\sigma \to \mathbf{E}); \vartheta_1)$$

$$= \sum_{\sigma \in \mathfrak{S}_r} \operatorname{sgn} \sigma \prod_{i=1}^r q^{i(A^{(i)} - A^{(\sigma(i))})} (GF(P(\mathcal{A}_{\sigma(i)} \to \mathcal{E}_i); \mathrm{maj})$$

$$- q^{i(2A^{(\sigma(i))} + D + 1)} GF(P(\mathfrak{R} \mathcal{A}_{\sigma(i)} \to \mathcal{E}_i); \mathrm{maj}))$$

$$= \sum_{\sigma \in \mathfrak{S}_r} \operatorname{sgn} \sigma \prod_{i=1}^r q^{i(A^{(i)} - A^{(\sigma(i))})} \left(\begin{bmatrix} E_1^{(i)} + E_2^{(i)} - D \\ E_1^{(i)} - A^{(\sigma(i))} - D \end{bmatrix} \right.$$

$$\left. - q^{i(2A^{(\sigma(i))} + D + 1)} \begin{bmatrix} E_1^{(i)} + E_2^{(i)} - D \\ E_2^{(i)} - A^{(\sigma(i))} - D - 1 \end{bmatrix} \right)$$

$$= \det_{1 \leq s, t \leq r} \left(q^{s(A^{(s)} - A^{(t)})} \begin{bmatrix} E_1^{(s)} + E_2^{(s)} - D \\ E_1^{(s)} - A^{(t)} - D \end{bmatrix} \right.$$

$$\left. - q^{s(2A^{(t)} + D + 1)} \begin{bmatrix} E_1^{(s)} + E_2^{(s)} - D \\ E_2^{(s)} - A^{(t)} - D - 1 \end{bmatrix} \right), \tag{3.2.9}$$

which settles (3.2.1). $\square$

Now we consider the case that the paths lie above the line $x = y$.

Theorem 6. *Let* $\mathcal{A}_i = (-A^{(i)}, A^{(i)} + D)$ *and* $\mathcal{E}_i = (E_1^{(i)}, E_2^{(i)})$, $i = 1, 2, \ldots, r$, *be lattice points in the integer lattice* $\mathbb{Z}^2$ *such that*

$$A^{(1)} < A^{(2)} < \cdots < A^{(r)}, \tag{3.2.10a}$$

$$E_1^{(1)} > E_1^{(2)} > \cdots > E_1^{(r)} \quad \text{and} \quad E_2^{(1)} \leq E_2^{(2)} \leq \cdots \leq E_2^{(r)} \tag{3.2.10b}$$

and

$$2A^{(i)} + D \geq 0 \quad \text{and} \quad E_1^{(i)} \leq E_2^{(i)}, \quad i = 1, 2, \ldots, r, \tag{3.2.10c}$$

hold. The generating function $\sum q^{\mathrm{maj}(\mathfrak{P})}$, *where the sum is over all nonintersecting families* $\mathfrak{P} = (P_1, \ldots, P_r)$ *of lattice paths which lie above the line* $x = y$, $P_i : \mathcal{A}_i \to \mathcal{E}_i$, $i = 1, 2, \ldots, r$, *and where* $\mathrm{maj}\,\mathfrak{P}$ *is given by (3.0.1), is equal to the expression*

$$\det_{1 \leq s, t \leq r} \left(q^{s(A^{(s)} - A^{(t)})} \left(\begin{bmatrix} E_1^{(s)} + E_2^{(s)} - D \\ E_2^{(s)} - A^{(t)} - D \end{bmatrix} \right. \right.$$

$$\left. \left. - q^{(s-1)(2A^{(t)} + D + 1)} \begin{bmatrix} E_1^{(s)} + E_2^{(s)} - D \\ E_1^{(s)} - A^{(t)} - D - 1 \end{bmatrix} \right) \right). \tag{3.2.11}$$

PROOF. The proof of Theorem 5 can be copied almost word-by-word. We only point out the differences.

Here a weight function ϑ_2 is introduced by

$$\vartheta_2(\mathfrak{P}) = (-1)^{\|\eta\|} \operatorname{sgn} \sigma \, q^{\sum_{i=1}^r \left(i(A^{(i)} - A^{(\sigma(i))}) + (i-1)\eta_i(2A^{(\sigma(i))} + D + 1) \right)} q^{\operatorname{maj} \mathfrak{P}}$$
$$\text{if } \mathfrak{P} \in Q(\mathfrak{R}^\eta \mathbf{A}_\sigma \to \mathbf{E}). \quad (3.2.12)$$

Note that the only difference with (3.2.2) is the term $(i-1)$ instead of i before the η_i.

What has to be determined, is the generating function $GF(\bar{Q}(\mathbf{A} \to \mathbf{E})^+, \vartheta_2)$, where $\bar{Q}(\mathbf{A} \to \mathbf{E})^+$ is the set of all families $\mathfrak{P} = (P_1, \ldots, P_r)$ of nonintersecting lattice paths which lie *above* $x = y$, $P_i : \mathcal{A}_i \to \mathcal{E}_i$. In order to do this, an involution Θ_2 is defined in the same way as Θ_1, only that (3.2.4d) has to be replaced by

$$P_1'' = \Phi^{(2)}(P_1), \quad (3.2.13)$$

where the map $\Phi^{(2)}$ is defined in Proposition 26.

The computations (3.2.6), (3.2.7), (3.2.9) run through quite analogously and can therefore be omitted. The only changes which have to be made are caused by the slightly different weight function ϑ_2 and that instead of (5.2.20) equation (5.2.22) has to be used. $\square$

There are again special cases where, by use of Lemma 34, the determinants in (3.2.1) and (3.2.11) can be evaluated. The result corresponding to Theorem 5 is given in Theorem 7. For a curious implication of Theorem 7 see Remark (4) at the end of section 4.

Theorem 7. Let $\mathcal{A}_i = (A^{(i)} + D, -A^{(i)})$ and $\mathcal{E}_i = (E+i, E+2-i)$, $i = 1, 2, \ldots, r$, be lattice points in the integer lattice $\mathbb{Z}^2$ such that (3.1.1a) and

$$2A^{(i)} + D \geq 0, \quad i = 1, 2, \ldots, r,$$

hold. The generating function $\sum q^{\operatorname{maj}(\mathfrak{P})}$ where the sum is over all nonintersecting families $\mathfrak{P} = (P_1, \ldots, P_r)$ of lattice paths which lie below the line $x = y$, $P_i : \mathcal{A}_i \to \mathcal{E}_i$, $i = 1, 2, \ldots, r$, is equal to the expression

$$\prod_{i=1}^r \frac{[2E + 2i - D]!}{[E + r - A^{(i)} - D]! \, [E + r + 1 + A^{(i)}]!}$$
$$\times \prod_{1 \leq i < j \leq r} [A^{(j)} - A^{(i)}] \prod_{1 \leq i \leq j \leq r} [A^{(i)} + A^{(j)} + D + 1] \,. \quad (3.2.14)$$

More generally, if the final points of the lattice paths are $\mathcal{E}_i = (E_1 + i, E_2 - i)$, $i = 1, 2, \ldots, r$, with $E_1 \geq E_2 - 2$, then the generating function $\sum q^{\operatorname{maj}(\mathfrak{P})}$ for all nonintersecting families $\mathfrak{P} = (P_1, \ldots, P_r)$ of lattice paths which lie below the line $x = y$, $P_i : \mathcal{A}_i \to \mathcal{E}_i$, $i = 1, 2, \ldots, r$, is equal to the expression

$$\sum_{\eta \in \{0,1\}^r} (-1)^{\|\eta\|} q^{\sum_{i=1}^r i\eta_i(2A^{(i)}+D+1)}$$

$$\prod_{i=1}^r \frac{[E_1 + E_2 + i - D - 1]!}{[E_2 + A^{(i)} - \eta_i(2A^{(i)} + D + 1) - 1]! \, [E_1 + r - D - A^{(i)} + \eta_i(2A^{(i)} + D + 1)]!}$$

$$\prod_{1 \le i < j \le r} [A^{(j)} - \eta_j(2A^{(j)} + D + 1) - A^{(i)} + \eta_i(2A^{(i)} + D + 1)] \quad (3.2.15)$$

PROOF. First we prove (3.2.14). If the final points of the lattice paths are $\mathcal{E}_i = (E+i, E+2-i)$, $i = 1, 2, \ldots, r$, then according to Theorem 5, the generating function in question is

$$\det_{1 \le s, t \le r} \left(q^{s(A^{(s)} - A^{(t)})} \left(\begin{bmatrix} 2E + 2 - D \\ E + s - A^{(t)} - D \end{bmatrix} - q^{s(2A^{(t)} + D + 1)} \begin{bmatrix} 2E + 2 - D \\ E - s - A^{(t)} - D + 1 \end{bmatrix} \right) \right). \quad (3.2.16)$$

Taking some factors out of the determinant yields that this is equal to

$$q^{\sum_{i=1}^r iA^{(i)}} \prod_{i=1}^r \frac{[2E + 2 - D]!}{[E + r - A^{(i)} - D]! \, [E + r + A^{(i)} + 1]!}$$

$$\times \det \Big(q^{-sA^{(t)}} [E + r - A^{(t)} - D] \cdots [E + s + 1 - A^{(t)} - D]$$

$$\cdot [E + r + A^{(t)} + 1] \cdots [E + s + 2 + A^{(t)}]$$

$$\cdot \big([E + s + 1 + A^{(t)}] \cdots [E - s + 3 + A^{(t)}]$$

$$- q^{s(2A^{(t)} + D + 1)} [E + s - A^{(t)} - D] \cdots [E - s + 2 - A^{(t)} - D] \big) \Big). \quad (3.2.17)$$

Setting $X_t = q^{A^{(t)} + D/2 + 1/2}$, $A_i = -q^{-E-i+D/2-1/2}$, the determinant in (3.2.17) can be written in the form

$$(-1)^{\binom{r}{2}} q^{2 \sum_{i=1}^r \sum_{j=i+1}^r (E+j-D/2+1/2)} q^{\sum_{i=1}^r \sum_{j=-i+2}^0 (E+j-D/2+1/2)}$$

$$\det \Big((A_r + X_t) \cdots (A_{s+1} + X_t)(A_r + 1/X_t) \cdots (A_{s+1} + 1/X_t)$$

$$\cdot q^{s(D/2+1/2)} \big((1/A_s + 1/X_t) \cdots (1/A_1 + 1/X_t)(A_0 + X_t) \cdots (A_{-s+2} + X_t)$$

$$- (1/A_s + X_t) \cdots (1/A_1 + X_t)(A_0 + 1/X_t) \cdots (A_{-s+2} + 1/X_t) \big) \Big). \quad (3.2.18)$$

This last determinant is almost in a form suitable for application of Lemma 34. The reader now should consult the very beginning of subsection 5.5 where the notions *Laurent polynomial* and *degree* of a Laurent polynomial are explained. Let us define

$$p_{i-1}(X) := q^{i(D/2+1/2)} \Big(\prod_{j=1}^{i}(1/A_j + 1/X) \prod_{j=-i+2}^{0} (A_j + X)$$

$$- \prod_{j=1}^{i}(1/A_j + X) \prod_{j=-i+2}^{0} (A_j + 1/X)\Big) \Big/ \Big(X - \frac{1}{X}\Big).$$

The denominator $X - 1/X$ in fact is a factor of the numerator of p_i, which is seen by setting $X = \pm 1$. Therefore $X - 1/X$ cancels out. Hence, the p_i's are Laurent polynomials of degree i. Moreover there obviously holds $p_i(1/X) = p_i(X)$. Taking $X_t - 1/X_t$ out of the t-th column of the determinant in (3.2.18), it is seen that Lemma 34 with $C = 1$ and the above choice of p_i can be applied to evaluate this determinant. Substitution of the result into (3.2.17) together with some cumbersome manipulations finally furnishes the expression (3.2.14).

Now we turn to the proof of (3.2.15). If the final points of the lattice paths are $\mathcal{E}_i = (E_1 + i, E_2 - i)$, $i = 1, 2, \ldots, r$, then according to Theorem 5, the generating function in question is

$$\det_{1 \le s,t \le r} \left(q^{s(A^{(s)} - A^{(t)})} \left(\begin{bmatrix} E_1 + E_2 - D \\ E_1 + s - A^{(t)} - D \end{bmatrix} \right.\right.$$

$$\left.\left. - q^{s(2A^{(t)}+D+1)} \begin{bmatrix} E_1 + E_2 - D \\ E_1 + s + A^{(t)} + 1 \end{bmatrix} \right) \right).$$

Next we use the linearity of the determinant in the columns to expand it as an alternating sum,

$$\sum_{\eta \in \{0,1\}^r} (-1)^{\|\eta\|} \det_{1 \le s,t \le r} \left(q^{s(A^{(s)} - A^{(t)}) + \eta_t s(2A^{(t)}+D+1)} \right.$$

$$\left. \begin{bmatrix} E_1 + E_2 - D \\ E_1 + s - A^{(t)} - D + \eta_t(2A^{(t)} + D + 1) \end{bmatrix} \right),$$

or equivalently,

$$\sum_{\eta \in \{0,1\}^r} (-1)^{\|\eta\|} q^{\sum_{i=1}^{r} i \eta_i (2A^{(i)}+D+1)} \det_{1 \le s,t \le r} \left(q^{s(A^{(s)} - \eta_s(2A^{(s)}+D+1) - A^{(t)} + \eta_t(2A^{(t)}+D+1))} \right.$$

$$\left. \begin{bmatrix} E_1 + E_2 - D \\ E_1 + s - A^{(t)} - D + \eta_t(2A^{(t)} + D + 1) \end{bmatrix} \right).$$

The determinants in this expression have already been evaluated. In the proof of Theorem 3, only $A^{(t)}$ has to be replaced by $A^{(t)} - \eta_t(2A^{(t)} + D + 1)$. The result is the expression (3.1.9), with this replacement. Thus we obtain (3.2.15). $\qquad\square$

The analogous specializations for Theorem 6 yield the following result.

Theorem 8. Let $\mathcal{A}_i = (-A^{(i)}, A^{(i)} + D)$ and $\mathcal{E}_i = (E+2-i, E+i)$, $i = 1, 2, \ldots, r$, be lattice points in the integer lattice $\mathbb{Z}^2$ such that the (3.2.10a) and

$$2A^{(i)} + D \geq 0, \quad i = 1, 2, \ldots, r,$$

hold. The generating function $\sum q^{\mathrm{maj}(\mathfrak{P})}$ where the sum is over all nonintersecting families $\mathfrak{P} = (P_1, \ldots, P_r)$ of lattice paths which lie above the line $x = y$, $P_i : \mathcal{A}_i \to \mathcal{E}_i$, $i = 1, 2, \ldots, r$ is equal to the expression

$$q^{\binom{r+1}{2}+r(E-D)-\sum_{i=1}^{r}A^{(i)}} \prod_{i=1}^{r} \frac{[2E + 2i - D]!}{[E + r - A^{(i)} - D]!\,[E + r + 1 + A^{(i)}]!}$$

$$\times \prod_{1 \leq i < j \leq r} [A^{(j)} - A^{(i)}] \prod_{1 \leq i \leq j \leq r} [A^{(i)} + A^{(j)} + D + 1] . \quad (3.2.19)$$

More generally, if the final points of the lattice paths are $\mathcal{E}_i = (E_1 - i, E_2 + i)$, $i = 1, 2, \ldots, r$, with $E_1 \leq E_2 + 2$, then the generating function $\sum q^{\mathrm{maj}(\mathfrak{P})}$ for all nonintersecting families $\mathfrak{P} = (P_1, \ldots, P_r)$ of lattice paths which lie above the line $x = y$, $P_i : \mathcal{A}_i \to \mathcal{E}_i$, $i = 1, 2, \ldots, r$, is equal to the expression

$$\sum_{\eta \in \{0,1\}^r} (-1)^{\|\eta\|} q^{\sum_{i=1}^{r}(i-1)\eta_i(2A^{(i)}+D+1)}$$

$$\prod_{i=1}^{r} \frac{[E_1 + E_2 + i - D - 1]!}{[E_1 + A^{(i)} - \eta_i(2A^{(i)} + D + 1) - 1]!\,[E_2 + r - D - A^{(i)} + \eta_i(2A^{(i)} + D + 1)]!}$$

$$\prod_{1 \leq i < j \leq r} [A^{(j)} - \eta_j(2A^{(j)} + D + 1) - A^{(i)} + \eta_i(2A^{(i)} + D + 1)] \quad (3.2.20)$$

PROOF. First we prove (3.2.19). According to Theorem 6, the generating function in question is

$$\det_{1 \leq s, t \leq r} \left(q^{s(A^{(s)} - A^{(t)})} \left(\begin{bmatrix} 2E + 2 - D \\ E + s - A^{(t)} - D \end{bmatrix} \right. \right.$$

$$\left. \left. - q^{(s-1)(2A^{(t)}+D+1)} \begin{bmatrix} 2E + 2 - D \\ E - s - A^{(t)} - D + 1 \end{bmatrix} \right) \right). \quad (3.2.21)$$

This is almost the same expression as that in (3.2.16). Hence, the computation proceeds quite analogously to that in the previous proof. Again, setting $X_t = q^{A^{(t)}+D/2+1/2}$, $A_i = -q^{-E-i+D/2-1/2}$, and taking some factors out of the determinant, we deduce that (3.2.21) can be written in the form

$$q^{\sum_{i=1}^{r} iA^{(i)}} \prod_{i=1}^{r} \frac{[2E+2-D]!}{[E+r-A^{(i)}-D]![E+r+A^{(i)}+1]!} q^{\sum_{i=1}^{r}(-A^{(i)}-D/2-1/2)}$$

$$(-1)^{\binom{r}{2}} q^{2\sum_{i=1}^{r}\sum_{j=i+1}^{r}(E+j-D/2+1/2)} q^{\sum_{i=1}^{r}\sum_{j=-i+2}^{1}(E+j-D/2+1/2)}$$

$$\det\Big((A_r+X_t)\cdots(A_{s+1}+X_t)(A_r+1/X_t)\cdots(A_{s+1}+1/X_t)$$

$$\cdot q^{s(D/2+1/2)}\big((1/A_s+1/X_t)\cdots(1/A_2+1/X_t)(A_1+X_t)\cdots(A_{-s+2}+X_t)$$

$$-(1/A_s+X_t)\cdots(1/A_2+X_t)(A_1+1/X_t)\cdots(A_{-s+2}+1/X_t)\big)\Big).$$

$$(3.2.22)$$

Here we have to choose

$$p_{i-1}(X) := q^{i(D/2+1/2)}\Big(\prod_{j=2}^{i}(1/A_j+1/X)\prod_{j=-i+2}^{1}(A_j+X)$$

$$-\prod_{j=2}^{i}(1/A_j+X)\prod_{j=-i+2}^{1}(A_j+1/X)\Big)\Big/\Big(X-\frac{1}{X}\Big).$$

Also these p_i's are Laurent polynomials of degree i which satisfy $p_i(1/X) = p_i(X)$. Again, after having taken $X_t - 1/X_t$ out of the determinant in (3.2.22), Lemma 34 with $C = 1$ and the above choice of p_i is applied to evaluate this determinant. This gives almost the same result as in the previous proof only that some power of q remains, which turns out to be $q^{\binom{r+1}{2}+r(E-D)-\sum_{i=1}^{r}A^{(i)}}$.

In order to prove (3.2.20), we use Theorem 6 with $E_1^{(i)} = E_1 - i$ and $E_2^{(i)} = E_2 + i$, $i = 1, 2, \ldots, r$. By (3.2.11) the generating function in question is

$$\det_{1\leq s,t\leq r}\Bigg(q^{s(A^{(s)}-A^{(t)})}\Bigg(\begin{bmatrix} E_1+E_2-D \\ E_2+s-A^{(t)}-D \end{bmatrix}$$

$$-q^{(s-1)(2A^{(t)}+D+1)}\begin{bmatrix} E_1+E_2-D \\ E_2+s+A^{(t)}+1 \end{bmatrix}\Bigg)\Bigg).$$

As in the proof of (3.2.15) we use the linearity of the determinant in the columns to expand it as an alternating sum,

$$\sum_{\eta\in\{0,1\}^r}(-1)^{\|\eta\|}\det_{1\leq s,t\leq r}\Bigg(q^{s(A^{(s)}-A^{(t)})+\eta_t s(2A^{(t)}+D+1)-\eta_t(2A^{(t)}+D+1)}$$

$$\begin{bmatrix} E_1+E_2-D \\ E_2+s-A^{(t)}-D+\eta_t(2A^{(t)}+D+1) \end{bmatrix}\Bigg),$$

or equivalently,

$$\sum_{\eta \in \{0,1\}^r} (-1)^{\|\eta\|} q^{\sum_{i=1}^r (i-1)\eta_i(2A^{(i)}+D+1)}$$

$$\det_{1 \le s, t \le r} \left(q^{s(A^{(s)}-\eta_s(2A^{(s)}+D+1)-A^{(t)}+\eta_t(2A^{(t)}+D+1))} \right.$$

$$\left. \begin{bmatrix} E_1 + E_2 - D \\ E_2 + s - A^{(t)} - D + \eta_t(2A^{(t)}+D+1) \end{bmatrix} \right).$$

The determinants in this expression are evaluated by using the computation in the proof of Theorem 3, with E_1 and E_2 interchanged and $A^{(t)}$ replaced by $A^{(t)}-\eta_t(2A^{(t)}+D+1)$. The result is the expression (3.1.9), with these replacements. Thus we obtain (3.2.20).   □

3.3. Tableaux generating functions. As first application we give a nice product formula for a generating function for pairs of tableaux with odd parts.

Theorem 9. *The generating function $\sum q^{n(\tau_1)+n(\tau_2)}$ for pairs (τ_1, τ_2) of tableaux of identical shape with at most r columns and with only odd parts which lie between 1 and $2n-1$ is given by*

$$\prod_{1 \le i,j \le n} \frac{[r+i+j-1]_{q^2}}{[i+j-1]_{q^2}} . \tag{3.3.1}$$

Equivalently, in terms of Schur functions,

$$\sum_{\lambda, \lambda_1 \le r} s_\lambda^2(q^{2n-1}, q^{2n-3}, \ldots, q) = \prod_{1 \le i,j \le n} \frac{[r+i+j-1]_{q^2}}{[i+j-1]_{q^2}} .$$

PROOF. By Proposition 29 the pairs of tableaux under consideration via the map Δ_2 are in one-to-one correspondence with families $\mathfrak{P} = (P_1, \ldots, P_r)$ of nonintersecting lattice paths, $P_i : (2i-2, -2i+2) \to (2n+2i-2, 2n-2i+2)$, where the North-East corners of the paths have odd coordinates. Moreover the relation (5.4.6) holds. Such families $\mathfrak{P}$ are modified in the following way. While fixing the starting and final points of the paths, shift all North-East corners of the paths of $\mathfrak{P}$ into the direction $(-1,1)$. More precisely, in each path the (horizontal) step before the first vertical step is turned into a vertical one, and the (vertical) step after the last horizontal step is turned into a horizontal one. Figure 3 should clarify what is meant.

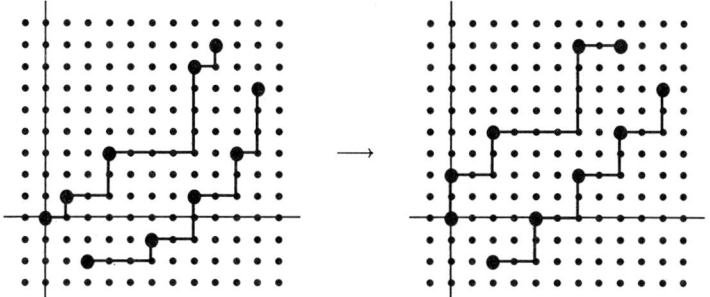

Figure 3

We obtain a new family $\bar{\mathfrak{P}} = (\bar{P}_1, \ldots, \bar{P}_r)$ of nonintersecting lattice paths consisting of double steps (i.e. only of the steps $(2, 0)$ and $(0, 2)$), $\bar{P}_i : (2i - 2, -2i + 2) \to (2n + 2i - 2, 2n - 2i + 2)$. Besides we obviously have $\text{maj}\,\mathfrak{P} = \text{maj}\,\bar{\mathfrak{P}}$. Therefore, in order to compute the desired generating function for pairs of tableaux, the generating function $\sum q^{\text{maj}\,\bar{\mathfrak{P}}}$ for nonintersecting families $\bar{\mathfrak{P}}$ of the above type has to be considered. But this generating function is evaluated using Theorem 3 with q replaced by q^2, $A^{(i)} = i - 1$, $E_1 = n - 1$, and $E_2 = n + 1$. With these choices of the parameters from (3.1.9) we obtain the expression

$$\prod_{i=1}^{r} \frac{[2n + i - 1]_{q^2}!}{[n + i - 1]_{q^2}! \, [n + r - i]_{q^2}!} \prod_{1 \le i < j \le r} [j - i]_{q^2} .$$

It is not difficult to convert this expression into (3.3.1). □

Our next application concerns a result about tableaux with even rows. It has been previously obtained by Désarménien [11, Théorème 1.2, first identity], Proctor [37, Theorem 1, case (CYI)], and Stembridge [40, Corollary 4.3 (a)].

Theorem 10. *The generating function for tableaux with rows of even length which do not exceed $2r$, and with parts between 1 and n, is given by*

$$\prod_{1 \le i \le j \le n} \frac{[2r + i + j]}{[i + j]} . \tag{3.3.2}$$

Equivalently, in terms of Schur functions,

$$\sum_{\lambda, \, \lambda_1 \le r} s_{2\lambda}(q^n, q^{n-1}, \ldots, q) = \prod_{1 \le i \le j \le n} \frac{[2r + i + j]}{[i + j]} .$$

PROOF. By the map Δ_3 defined in Proposition 30 the tableaux under consideration uniquely correspond to families $\mathfrak{P} = (P_1, \ldots, P_r)$ of nonintersecting lattice paths which lie below the line $x = y$, $P_i : (i - 1, -i + 1) \to (n + i, n + 2 - i)$. Moreover the relation (5.4.7) is satisfied. Hence, if in Theorem 7 we set $A^{(i)} = i - 1$, $D = 0$, and $E = n$, we obtain for the generating function in question the expression

$$\prod_{i=1}^{r} \frac{[2n+2i]!}{[n+r-i+1]!\,[n+r+i]!} \prod_{1\le i<j\le r} [j-i] \prod_{1\le i\le j\le r} [i+j-1].$$

It is easy to verify that this expression simplifies to (3.3.2). $\square$

Our next theorem refines a result about tableaux with even rows and odd parts that was also previously obtained by Désarménien [11, Théoréme 1.2, second identity], Proctor [37, Theorem 1, case (CYH)], and Stembridge [40, Corollary 4.3 (b)]. Their result is the $p = 0$ special case of Theorem 11. The $c = 2r + 1$ case of Theorem 11 was first found by Désarménien [12, Théoréme 2, second identity].

Theorem 11. *The generating function for tableaux with p odd rows, with at most c columns, and with only odd parts which lie between 1 and $2n - 1$, is given by*

$$q^{p^2} \frac{[2r+2p]_{q^2}\,[r]_{q^2}}{[2r+p]_{q^2}\,[r+p]_{q^2}} \begin{bmatrix} n \\ p \end{bmatrix}_{q^2} \frac{\begin{bmatrix} n+2r \\ n \end{bmatrix}_{q^2}}{\begin{bmatrix} n+2r+p \\ n \end{bmatrix}_{q^2}}$$

$$\times \prod_{i=1}^{n} \frac{[r+i]_{q^2}}{[i]_{q^2}} \prod_{1\le i<j\le n} \frac{[2r+i+j]_{q^2}}{[i+j]_{q^2}} \qquad \text{if } c = 2r \quad (3.3.3a)$$

and

$$q^{p^2} \begin{bmatrix} n \\ p \end{bmatrix}_{q^2} \prod_{i=1}^{n} \frac{[r+i]_{q^2}}{[i]_{q^2}} \prod_{1\le i<j\le n} \frac{[2r+i+j]_{q^2}}{[i+j]_{q^2}} \qquad \text{if } c = 2r+1. \quad (3.3.3b)$$

The formulation of this result in terms of Schur functions is that the sum $\sum s_\lambda(q^{2n-1}, q^{2n-3}, \ldots, q)$, where the sum is over all partitions λ with exactly p odd parts, and where all the parts do not exceed c, is given by the expressions in (3.3.3).

PROOF. We have to distinguish between c being even and c being odd.

First let $c = 2r$. By Proposition 33 we know that, via the map Δ_6, the tableaux under consideration are in one-to-one correspondence with pairs $(\mathfrak{P}, S)$, where $\mathfrak{P} = (P_1, \ldots, P_r)$ is a family of nonintersecting lattice paths which lie below the line $x = y$, $P_i : (2i-2, -2i+2) \to (2n+2i-2, 2n-2i+2)$, where the North-East corners of all the paths have odd coordinates, with the property that if P_r starts with $2h+1$ horizontal steps followed by a vertical step then S is a p-subset of $\{1, 2, \ldots, 2h - 1\}$. For a path which starts with H horizontal steps followed by a vertical one, we say that it *first turns after H steps*. Moreover, Δ_6 satisfies the relation (5.4.14). We claim that these pairs $(\mathfrak{P}, S)$ bijectively correspond to families $\hat{\mathfrak{P}} = (\hat{P}_1, \ldots, \hat{P}_r)$ of nonintersecting lattice paths consisting only of double steps, $\hat{P}_i : (2i, -2i+2) \to (2n+2i, 2n+4-2i)$,

$i = 1, \ldots, r - 1$, $\hat{P}_r : (2r + 2p, -2r - 2p + 2) \to (2n + 2r, 2n + 4 - 2r)$, which lie below $x = y$, such that

$$\text{maj } \mathfrak{P} + \|S\| = \text{maj } \hat{\mathfrak{P}} + p^2. \qquad (3.3.4)$$

Suppose that this had been done. The generating function $\sum_{\hat{\mathfrak{P}}} q^{\text{maj } \hat{\mathfrak{P}}}$ for this last set of families of nonintersecting lattice paths is obtained from Theorem 7 by replacing q by q^2 and then setting $A^{(i)} = i - 1$, $i = 1, \ldots, r - 1$, $A^{(r)} = r + p - 1$, $D = 1$, $E = n$. The resulting expression because of (3.3.4) has to be multiplied by q^{p^2}. We thus get

$$q^{p^2} \prod_{i=1}^{r} \frac{[2n + 2i - 1]_{q^2}!}{[n + r - i]_{q^2}! \, [n + r + i]_{q^2}!} \cdot \frac{[n]_{q^2}! \, [n + 2r]_{q^2}!}{[n - p]_{q^2}! \, [n + 2r + p]_{q^2}!}$$

$$\prod_{1 \le i < j \le r} [j - i]_{q^2} \prod_{i=1}^{r-1} \frac{[r + p - i]_{q^2}}{[r - i]_{q^2}}$$

$$\prod_{1 \le i \le j \le r} [i + j]_{q^2} \prod_{i=1}^{r} \frac{[r + p + i]_{q^2}}{[r + i]_{q^2}} \cdot \frac{[2r + 2p]_{q^2}}{[2r + p]_{q^2}},$$

which after some manipulation is turned into (3.3.3a).

Thus it remains to construct this bijection. Let $(\mathfrak{P}, S)$ be a pair as described above. First we modify $\mathfrak{P}$ by performing the same operation on $\mathfrak{P}$ as in the proof of Theorem 9. Namely, while fixing the starting and final points of the paths, shift all North-East corners of the paths of $\mathfrak{P}$ into the direction $(-1, 1)$. More precisely, in each path the (horizontal) step before the first vertical step is turned into a vertical one, and the (vertical) step after the last horizontal step is turned into a horizontal one (cf. Figure 3). We thus obtain a new family $\bar{\mathfrak{P}} = (\bar{P}_1, \ldots, \bar{P}_r)$ of nonintersecting lattice paths consisting only of double steps, $\bar{P}_i : (2i - 2, -2i + 2) \to (2n + 2i - 2, 2n - 2i + 2)$, which lie below the line $x = y - 2$. Besides, $\bar{P}_r$ first turns after $2h$ steps. By shifting the family $\bar{\mathfrak{P}}$ in the direction $(2, 0)$ and attaching a double step at the end of each path, we obtain the family $\tilde{\mathfrak{P}} = (\tilde{P}_1, \ldots, \tilde{P}_r)$ of nonintersecting lattice paths consisting only of double steps, $\tilde{P}_1 : (2i, -2i + 2) \to (2n + 2i, 2n + 4 - 2i)$, which lie below $x = y$. Besides, $\tilde{P}_r$ first turns after $2h$ steps. We remark that the added double step at the end of each path in fact is superflous since, because of the location of the final points and because of the requirements of being nonintersecting and lying below $x = y$, each path of $\tilde{\mathfrak{P}}$ has to end with a vertical double step. These double steps only are inserted to fit with Theorem 7.

Now we come to the set S. Using (5.1.2) with q replaced by q^2 it is easily seen that the generating function $\sum q^{\|S\|}$ for the p-subsets of $\{1, 3, \ldots, 2h - 1\}$ is given by

$$\sum_{S \subseteq \{1, 3, \ldots, 2h-1\}, \, |S| = p} q^{\|S\|} = q^{p^2} \begin{bmatrix} h \\ p \end{bmatrix}_{q^2}. \qquad (3.3.5)$$

On the other hand, by replacing q by q^2 in (5.1.13) we get that the major generating function for paths from $(2r + 2p, -2r - 2p + 2)$ to $(2r + 2h, -2r + 2)$ consisting only of double steps is given by

$$\sum_{\substack{P:(2r+2p,-2r-2p+2)\to(2r+2h,-2r+2)\\ P \text{ consists of double steps}}} q^{\operatorname{maj} P} = \begin{bmatrix} h \\ p \end{bmatrix}_{q^2}. \qquad (3.3.6)$$

A comparison of (3.3.6) and (3.3.5) shows that there is a bijection δ_1 between p-subsets S of $\{1,3,\dots,2h-1\}$ and paths P from $(2r+2p,-2r-2p+2)$ to $(2r+2h,-2r+2)$, consisting only of double steps, such that the relation

$$\|S\| = \operatorname{maj} \delta_1(S) + p^2 \qquad (3.3.7)$$

holds. (An explicit bijection could be extracted from [20].) Since $\tilde{P}_r$ starts at $(2r,-2r+2)$ and first turns after $2h$ steps, the point $(2r+2h,-2r+2)$ lies on $\tilde{P}_r$. Now we form a new path $\hat{P}_r$ by joining $\delta_1(S)$ and the portion of $\tilde{P}_r$ from $(2r+2h,-2r+2)$ to its final point $(2n+2r,2n+4-2r)$. Moreover, since the step after the point $(2r+2h,-2r+2)$ is a vertical one, we always are able to recover $\delta_1(S)$, and hence S, from $\hat{P}_r$. Summarizing, starting with the pair $(\mathfrak{P},S)$ we obtained the family $\hat{\mathfrak{P}} = (\tilde{P}_1,\dots,\tilde{P}_{r-1},\hat{P}_r)$. Besides, (3.3.4) holds. This gives the claimed bijection.

Finally, let $c = 2r+1$. We claim that the tableaux τ in question are in bijection with pairs (τ_e,S), where τ_e is a tableau with even rows, with at most $2r$ columns, and with only odd parts which lie between 1 and $2n-1$, and where S is a p-subset of $\{1,3,\dots,2n-1\}$. This is seen using Knuth's procedures ROW-DELETE and ROW-INSERT (see subsection 5.4). Given a tableau τ with p odd rows, perform the procedure ROW-DELETE for each of the odd rows, beginning from the lowest and working upward to the highest row. We thus obtain a sequence $n_1,n_2,\dots,n_p$ of odd integers, the n_1 being deleted by the first application of ROW-DELETE, etc. Clearly, $S := \{n_1,n_2,\dots,n_p\}$ is a subset of $\{1,3,\dots,2n-1\}$. Moreover we are left with a tableau τ_e without odd rows. The observation that the integers $n_1,n_2,\dots,n_p$ are in increasing order shows that by successively inserting $n_p,n_{p-1},\dots,n_1$ into τ_e utilizing ROW-INSERT the above procedure could be inverted. Hence it is a bijection which in addition satisfies $n(\tau) = n(\tau_e) + \|S\|$. Use of (3.3.3a) with $p = 0$ and of identity (3.3.5) with h replaced by n establishes (3.3.3b). $\square$

Theorem 11 also has a formulation in terms of symmetric plane partitions. A plane partition τ is called *symmetric* if $\tau_{ij} = \tau_{ji}$ for all i,j.

Corollary 12. *The generating function for symmetric plane partitions with at most n rows, with parts between 1 and c, and with exactly p odd entries on the main diagonal, is given by the expressions in (3.3.3).*

PROOF. By Theorem 11 the expressions in (3.3.3) are equal to the sum $\sum s_\lambda(q^{2n-1}, q^{2n-3},\dots,q)$ summed over all partitions λ with exactly p odd parts and $\lambda_1 \le c$. Examining the correspondence in [40, p. 497] more closely, it is easily verified that it is also a weight preserving-bijection between symmetric plane partitions with at most

n rows, with parts between 1 and c, and with exactly p odd entries on the main diagonal, and column-strict plane partitions with p odd rows, with at most c columns, and with parts being odd and between 1 and $2n - 1$. But the generating function for this set of column-strict plane partitions is equal to the above summation of Schur functions.   □

Now we are in the position to give a new proof of the MacMahon conjecture about symmetric plane partitions. The first proofs were given by Andrews [2], Macdonald [33, Ex. 16 and 17, pp. 51/52], and Proctor [36, Proposition 7.3].

Theorem 13. *The generating function for symmetric plane partitions with at most n rows and with parts between 1 and c is equal to*

$$\prod_{i=1}^{n} \frac{[c + 2i - 1]_q}{[2i - 1]_q} \prod_{1 \le i < j \le n} \frac{[c + i + j - 1]_{q^2}}{[i + j - 1]_{q^2}} . \tag{3.3.8}$$

Equivalently, the generating function for tableaux with at most c columns and with only odd parts which lie between 1 and $2n - 1$ is also given by (3.3.8). Or in terms of Schur functions: The sum $\sum_{\lambda, \lambda_1 \le c} s_\lambda(q^{2n-1}, q^{2n-3}, \ldots, q)$ is also equal to (3.3.8).

PROOF. We have to sum the expressions in (3.3.3) with respect to p. If the sum over all p of the expressions (3.3.3a) is converted into hypergeometric notation (cf. subsection 2.1), it reads

$$\prod_{i=1}^{n} \frac{[r + i]_{q^2}}{[i]_{q^2}} \prod_{1 \le i < j \le n} \frac{[2r + i + j]_{q^2}}{[i + j]_{q^2}}$$

$$\times {}_6\phi_5 \left[\begin{array}{c} q^{4r}, q^{2r+2}, -q^{2r+2}, q^{2r}, -q^{2r+1}, q^{-2n} \\ q^{2r}, -q^{2r}, q^{2r+2}, -q^{2r+1}, q^{4r+2n+2} \end{array} ; q^2, -q^{2n+1} \right] .$$

The ${}_6\phi_5$ can be evaluated by using the very well-poised ${}_6\phi_5$-summation (see [14, Appendix (II.21)])

$${}_6\phi_5 \left[\begin{array}{c} a, q\sqrt{a}, -q\sqrt{a}, b, c, q^{-n} \\ \sqrt{a}, -\sqrt{a}, aq/b, aq/c, aq^{n+1} \end{array} ; q, \frac{aq^{n+1}}{bc} \right] = \frac{(aq)_n (aq/bc)_n}{(aq/b)_n (aq/c)_n} . \tag{3.3.9}$$

Thus we obtain

$$\prod_{i=1}^{n} \frac{[r + i]_{q^2}}{[i]_{q^2}} \prod_{1 \le i < j \le n} \frac{[2r + i + j]_{q^2}}{[i + j]_{q^2}} \frac{(q^{4r+2}; q^2)_n (-q; q^2)_n}{(q^{2r+2}; q^2)_n (-q^{2r+1}; q^2)_n}$$

which after some manipulation yields (3.3.8) with $c = 2r$.

On the other hand, the sum over all p of the expressions (3.3.3b) is done by using the q-binomial theorem (see e.g. [1, (3.3.7)])

$$\sum_{k=0}^{N} q^{\binom{k}{2}} \begin{bmatrix} N \\ k \end{bmatrix} z^k = (1 + z)(1 + qz) \cdots (1 + q^{N-1}z) \tag{3.3.10}$$

with q replaced by q^2, $N = n$, and $z = q$. This gives

$$\prod_{i=1}^{n}(1 + q^{2i-1}) \prod_{i=1}^{n} \frac{[r+i]_{q^2}}{[i]_{q^2}} \prod_{1 \le i < j \le n} \frac{[2r+i+j]_{q^2}}{[i+j]_{q^2}}.$$

A little bit of manipulation shows that this expression is the same as that in (3.3.8) with $c = 2r + 1$.   □

REMARK. It is interesting to note that by the deleting-inserting procedure described at the end of the proof of Theorem 11, the theorem of Désarménien, Proctor, and Stembridge about tableaux with even rows and odd parts (Theorem 10 with $p = 0$) is equivalent with the MacMahon conjecture (Theorem 13) for $c = 2r + 1$. So in fact we could say that this result about tableaux with even rows and odd parts is a special case of the MacMahon conjecture. This equivalence was previously observed by Désarménien [12] although he is arguing by the Littlewood–Richardson rule.

IV. Counting by Strange Major Index

In this section we consider two generalizations of the major index. Remember (see subsection 2.1) that the major index of a path can be formulated in terms of contributions of the North-East corners of the path. The idea for our extensions of the major index is that every North-East corner which lies strictly above a fixed horizontal line contributes an extra weight, or every North-East corner which lies to the right-hand side of a fixed vertical line contributes an extra weight. We denote the statistics which depends on the y-coordinates of the North-East corners by ymaj and the statistics which depends on the x-coordinates of the North-East corners by xmaj. Let β be some real number, γ be an integer, and P be a path from $\mathcal{A} = (A_1, A_2)$ to $\mathcal{E} = (E_1, E_2)$. For $\gamma \geq A_2$ we introduce the statistics $\mathrm{ymaj}_{\beta;\gamma}$ by

$$\mathrm{ymaj}_{\beta;\gamma}(P) = \mathrm{maj}\, P + \beta \cdot |\{(p_1, p_2) : (p_1, p_2) \in NE(P) \text{ and } p_2 > \gamma\}|, \qquad \text{for } \gamma \geq A_2, \tag{4.0.1}$$

where $NE(P)$ denotes the set of North-East corners of P. For $\gamma \geq A_1$ we introduce the statistics $\mathrm{xmaj}_{\beta;\gamma}$ by

$$\mathrm{xmaj}_{\beta;\gamma}(P) = \mathrm{maj}\, P + \beta \cdot |\{(p_1, p_2) : (p_1, p_2) \in NE(P) \text{ and } p_1 \geq \gamma\}|, \qquad \text{for } \gamma \geq A_1. \tag{4.0.2}$$

For example for the path P_0 in our example in Figure 1 we have $\mathrm{ymaj}_{\beta;2}(P_0) = 16 + 2\beta$ or $\mathrm{ymaj}_{\beta;3}(P_0) = 16 + \beta$, and $\mathrm{xmaj}_{\beta;1}(P_0) = 16 + 3\beta$ or $\mathrm{xmaj}_{\beta;6}(P_0) = 16$. These two statistics $\mathrm{ymaj}_{\beta;\gamma}$ and $\mathrm{xmaj}_{\beta;\gamma}$ in the sequel will be referred to as *strange major indices*.

The definition (4.0.1) would not be reasonable for $\gamma < A_2$, and definition (4.0.2) would not be reasonable for $\gamma < A_1$, since in these cases all North-East corners of P would automatically contribute a β. But the observation (5.1.10), together with the remarks after this identity, suggests a useful extension for $\mathrm{ymaj}_{\beta;\gamma}$ and $\mathrm{xmaj}_{\beta;\gamma}$ for these values of γ. We define

$$\mathrm{ymaj}_{\beta;\gamma}(P) = \mathrm{xmaj}_{\beta;A_1 + A_2 - \gamma}(P) + \beta(A_2 - \gamma) \qquad \text{for } \gamma < A_2, \tag{4.0.3}$$

and

$$\mathrm{xmaj}_{\beta;\gamma}(P) = \mathrm{ymaj}_{\beta;A_1 + A_2 - \gamma}(P) + \beta(A_1 - \gamma) \qquad \text{for } \gamma < A_1. \tag{4.0.4}$$

From the definitions (4.0.1)–(4.0.4) we infer that for *all* integers γ there holds the relation

$$\mathrm{ymaj}_{\beta;\gamma}(P) = \mathrm{xmaj}_{\beta;A_1 + A_2 - \gamma}(P) + \beta(A_2 - \gamma). \tag{4.0.5}$$

Since in section 5 we shall most of the time use the array representation of paths, we also have to present formulas for computing the major indices of a path $P : \mathcal{A} \to \mathcal{E}$ in terms of its array representation

$$\begin{array}{ccccccc} A_1 \leq & a_1 & a_2 & \dots & a_k & \leq E_1 - 1 \\ A_2 + 1 \leq & b_1 & b_2 & \dots & b_k & \leq E_2 \end{array} \quad .$$

We have

$$
\mathrm{ymaj}_{\beta;\gamma}\, P = \begin{cases} \|a\| + \|b\| + \beta \cdot |\{t : b_t > \gamma\}| - k(A_1 + A_2) & \text{if } \gamma \geq A_2 \\[2mm] \|a\| + \|b\| + \beta \cdot |\{t : a_t \geq A_1 + A_2 - \gamma\}| & \\ \qquad + \beta(A_2 - \gamma) - k(A_1 + A_2) & \text{if } \gamma < A_2 \end{cases} \tag{4.0.6}
$$

and

$$
\mathrm{xmaj}_{\beta;\gamma}\, P = \begin{cases} \|a\| + \|b\| + \beta \cdot |\{t : a_t \geq \gamma\}| - k(A_1 + A_2) & \text{if } \gamma \geq A_1 \\[2mm] \|a\| + \|b\| + \beta \cdot |\{t : b_t > A_1 + A_2 - \gamma\}| & \\ \qquad + \beta(A_1 - \gamma) - k(A_1 + A_2) & \text{if } \gamma < A_1 \end{cases} \tag{4.0.7}
$$

These definitions admittedly seem to be rather artificial. The motivation of doing strange major counting is revealed in the proof of Theorem 21, where the Choi/Gouyou-Beauchamps problem which was mentioned in the introduction is solved by transferring it, with the help of Proposition 32, to a strange major counting problem for nonintersecting lattice paths.

Intending to count families $\mathfrak{P} = (P_1, \ldots, P_r)$ of nonintersecting lattice paths by strange major index, one could try the most general way to do that, namely to count P_i by $\mathrm{ymaj}_{\beta_i;\gamma_i}$ or $\mathrm{xmaj}_{\beta_i;\gamma_i}$, $i = 1, 2, \ldots, r$. But since we are going to rely on Proposition 27, in view of (5.3.2) and (5.3.3), we see that it can only be possible to analogize the procedures of subsection 2.2 if all the β_i's are equal and the γ_i's are successive integers. This motivates the following extension of the strange major indices to families $\mathfrak{P} = (P_1, \ldots, P_r)$, $P_i : \mathcal{A}_i \to \mathcal{E}_i$, $i = 1, 2, \ldots, r$, satisfying (2.2.2). We set

$$
\mathrm{ymaj}_{\beta;\gamma}(\mathfrak{P}) = \sum_{i=1}^{r} \mathrm{ymaj}_{\beta;\gamma-i+1}(P_i), \tag{4.0.8}
$$

and

$$
\mathrm{xmaj}_{\beta;\gamma}(\mathfrak{P}) = \sum_{i=1}^{r} \mathrm{xmaj}_{\beta;\gamma+i-r}(P_i). \tag{4.0.9}
$$

We remark that the effect of requiring (2.2.2) in these definitions is that in (4.0.8) the "highest" path is counted by $\mathrm{ymaj}_{\beta;\gamma}$, while in (4.0.9) the "lowest" path is counted by $\mathrm{xmaj}_{\beta;\gamma}$.

4.1. Without restrictions. We proceed in analogy with subsection 3.1. We start with the result for the strange major counting of unrestricted families of nonintersecting lattice paths with given starting and final points. In fact there are two results, one for ymaj and one for xmaj.

Theorem 14. *Let* $\mathcal{A}_i = (A^{(i)} + D, -A^{(i)})$ *and* $\mathcal{E}_i = (E_1^{(i)}, E_2^{(i)})$, $i = 1, 2, \ldots, r$, *be lattice points in the integer lattice* $\mathbb{Z}^2$ *such that (3.1.1) is satisfied.*

If γ *is an integer satisfying*

$$
D - E_1^{(1)} \leq \gamma \leq \min_{1 \leq i \leq r}(E_2^{(i)} + i - 1) \tag{4.1.1}
$$

then the generating function $\sum q^{\mathrm{ymaj}_{\beta;\gamma}}(\mathfrak{P})$, where the sum is over all nonintersecting families $\mathfrak{P} = (P_1, \ldots, P_r)$ of lattice paths, $P_i : \mathcal{A}_i \to \mathcal{E}_i$, $i = 1, 2, \ldots, r$, and where $\mathrm{ymaj}_{\beta;\gamma}(\mathfrak{P})$ is given by (4.0.8), is equal to the expression

$$\det_{1 \leq s,t \leq r} \left(q^{s(A^{(s)} - A^{(t)})} \sum_{j \geq 0} q^{j(j+\beta+\gamma+A^{(t)}-s+1)} \begin{bmatrix} -\beta \\ j \end{bmatrix} \begin{bmatrix} \beta + E_1^{(s)} + E_2^{(s)} - D \\ E_1^{(s)} - A^{(t)} - D - j \end{bmatrix} \right).$$

(4.1.2)

If γ is an integer satisfying

$$\max_{1 \leq i \leq r} \left(D - E_2^{(i)} + r - i \right) \leq \gamma \leq E_1^{(1)} + r - 1 \tag{4.1.3}$$

then the generating function $\sum q^{\mathrm{xmaj}_{\beta;\gamma}}(\mathfrak{P})$, summed over the same family of nonintersecting lattice paths, equals

$$\det_{1 \leq s,t \leq r} \left(q^{s(A^{(s)} - A^{(t)})} \sum_{j \geq 0} q^{j(j+\beta+\gamma-A^{(t)}+s-r)} \begin{bmatrix} -\beta \\ j \end{bmatrix} \begin{bmatrix} \beta + E_1^{(s)} + E_2^{(s)} - D \\ E_2^{(s)} + A^{(t)} - j \end{bmatrix} \right).$$

(4.1.4)

PROOF. The proof of Theorem 2 can be copied almost word-by-word. We only point out the differences. For the proof of (4.1.2), in (3.1.3) maj has to be replaced by $\mathrm{ymaj}_{\beta;\gamma}$, while for the proof of (4.1.4) it has to be replaced by $\mathrm{xmaj}_{\beta;\gamma}$. The computation (3.1.6) then also works because instead of (5.3.4) we can use (5.3.2) or (5.3.3), respectively. Finally, in the computation (3.1.8) maj has to be replaced by $\mathrm{ymaj}_{\beta;\gamma-i+1}$, respectively $\mathrm{xmaj}_{\beta;\gamma+i-r}$. Subsequently, the generating functions $GF(P(\mathcal{A}_{\sigma(i)} \to \mathcal{E}_i); \mathrm{ymaj}_{\beta;\gamma-i+1})$ and $GF(P(\mathcal{A}_{\sigma(i)} \to \mathcal{E}_i); \mathrm{xmaj}_{\beta;\gamma+i-r})$ are obtained using (5.1.11) and (5.1.12), respectively, instead of (5.1.13). It should be noted that (4.1.1) and (4.1.3) guarantee that the requirements of Lemma 23 are indeed satisfied. □

In the same special case as that in Theorem 3 we can get rid of the determinant. But, unfortunately, what remains is a multiple sum which evidently cannot be summed into a closed form (see Remark (3) at the end of this section), except of course that for $\beta = 0$ we obtain Theorem 3 again.

Theorem 15. Let $\mathcal{A}_i = (A^{(i)} + D, -A^{(i)})$ and $\mathcal{E}_i = (E_1 + i, E_2 - i)$, $i = 1, 2, \ldots, r$, be lattice points in the integer lattice $\mathbb{Z}^2$ such that (3.1.1a) holds.

If γ is an integer satisfying

$$D - E_1 - 1 \leq \gamma \leq E_2 - 1,$$

then the generating function $\sum q^{\mathrm{ymaj}_{\beta;\gamma}}(\mathfrak{P})$ where the sum is over all nonintersecting families $\mathfrak{P} = (P_1, \ldots, P_r)$ of lattice paths, $P_i : \mathcal{A}_i \to \mathcal{E}_i$, $i = 1, 2, \ldots, r$, is equal to the expression

$$\sum_{k_1, \ldots, k_r \geq 0} \left(\prod_{i=1}^r q^{k_i(k_i + \beta + \gamma + A^{(i)} - i + 1)} \begin{bmatrix} -\beta \\ k_i \end{bmatrix} \right.$$

$$\times \frac{[\beta + E_1 + E_2 + i - D - 1]!}{[\beta + E_2 + A^{(i)} + k_i - 1]! \, [E_1 + r - D - A^{(i)} - k_i]!} \prod_{1 \leq i < j \leq r} [A^{(j)} + k_j - A^{(i)} - k_i] \right).$$

(4.1.5a)

An alternative expression (which usually is computationally better) is

$$\sum_{k_1,\ldots,k_r \geq 0} \left(\prod_{i=1}^{r} q^{k_i(k_i+\beta+\gamma+A^{(i)}-i+1)} \begin{bmatrix} E_2 - \gamma - 1 \\ k_i \end{bmatrix} \right.$$

$$\left. \times \frac{[E_1 - D + \gamma + i]!}{[E_1 + r - D - A^{(i)} - k_i]! \, [\gamma + A^{(i)} + k_i]!} \prod_{1 \leq i < j \leq r} [A^{(j)} + k_j - A^{(i)} - k_i] \right).$$

$$(4.1.5b)$$

If γ is an integer satisfying

$$D - E_2 + r \leq \gamma \leq E_1 + r,$$

then the generating function $\sum q^{\mathrm{xmaj}_{\beta;\gamma}(\mathfrak{P})}$ for the same set of families $\mathfrak{P}$ of noninter-secting lattice paths equals

$$\sum_{k_1,\ldots,k_r \geq 0} \left(\prod_{i=1}^{r} q^{k_i(k_i+\beta+\gamma-A^{(i)}+i-r)} \begin{bmatrix} -\beta \\ k_i \end{bmatrix} \right.$$

$$\left. \times \frac{[\beta + E_1 + E_2 + i - D - 1]!}{[E_2 + A^{(i)} - k_i - 1]! \, [\beta + E_1 + r - D - A^{(i)} + k_i]!} \prod_{1 \leq i < j \leq r} [A^{(j)} - k_j - A^{(i)} + k_i] \right).$$

$$(4.1.6a)$$

An alternative expression is

$$\sum_{k_1,\ldots,k_r \geq 0} \left(\prod_{i=1}^{r} q^{k_i(k_i+\beta+\gamma-A^{(i)}+i-r)} \begin{bmatrix} E_1 - D + r - \gamma \\ k_i \end{bmatrix} \right.$$

$$\left. \times \frac{[E_2 + \gamma - i]!}{[E_2 + A^{(i)} - k_i - 1]! \, [\gamma - A^{(i)} + k_i]!} \prod_{1 \leq i < j \leq r} [A^{(j)} - k_j - A^{(i)} + k_i] \right).$$

$$(4.1.6b)$$

PROOF. For the $\mathrm{ymaj}_{\beta;\gamma}$-generating function set $E_1^{(i)} = E_1 + i$ and $E_2^{(i)} = E_2 - i$ in (4.1.2). This provides a determinant with a sum in each entry. The summation indices are independent, but of course it is allowed to make them dependent by setting all the summation indices in the t-th column equal to k_t, say. Then we use the linearity in the t-th column to obtain for the $\mathrm{ymaj}_{\beta;\gamma}$-generating function the expression

$$\sum_{k_1,\ldots,k_r \geq 0} \prod_{i=1}^{r} q^{k_i(k_i+\beta+\gamma+A^{(i)}-i+1)} \begin{bmatrix} -\beta \\ k_i \end{bmatrix}$$

$$\times \det \left(q^{s(A^{(s)}+k_s-A^{(t)}-k_t)} \begin{bmatrix} \beta + E_1 + E_2 - D \\ E_1 + s - D - A^{(t)} - k_t \end{bmatrix} \right).$$

The determinant in this last expression has already been evaluated. In the proof of Theorem 3, E_2 has to be replaced by $E_2 + \beta$ and $A^{(t)}$ by $A^{(t)} + k_t$. The result is the expression in (3.1.9), with these replacements. This furnishes (4.1.5a).

The expression (4.1.5b) follows from Gustafson's [22] recently discovered A_r-type extension of Heine's transformation (5.1.6),

$$
\sum_{k_1,\ldots,k_r \geq 0} \left(\prod_{i=1}^{r} q^{k_i(1-i)} Z^{k_i} \frac{(A)_{k_i}(BX_i)_{k_i}}{(q)_{k_i}(CX_i)_{k_i}} \right) \prod_{1 \leq i < j \leq r} \frac{1 - \frac{X_j}{X_i} q^{k_j - k_i}}{1 - \frac{X_j}{X_i}}
$$
$$
= \prod_{i=1}^{r} \frac{(Cq^{i-r}/B)_\infty (BZX_i)_\infty}{(Zq^{i-r})_\infty (CX_i)_\infty} \sum_{k_1,\ldots,k_r \geq 0} \left(\prod_{i=1}^{r} q^{k_i(1-i)} \left(\frac{C}{B} \right)^{k_i} \right.
$$
$$
\left. \frac{(ABZ/C)_{k_i}(BX_i)_{k_i}}{(q)_{k_i}(BZX_i)_{k_i}} \right) \prod_{1 \leq i < j \leq r} \frac{1 - \frac{X_j}{X_i} q^{k_j - k_i}}{1 - \frac{X_j}{X_i}}. \quad (4.1.7)
$$

The proofs of (4.1.6a,b) are similar. $\quad \square$

4.2. With a diagonal boundary. Here we proceed in analogy with subsection 3.2. First we state the strange major counting result for families of nonintersecting lattice paths which lie below the line $x = y$. There is only a result for ymaj but none for xmaj.

Theorem 16. *Let $\mathcal{A}_i = (A^{(i)} + D, -A^{(i)})$ and $\mathcal{E}_i = (E_1^{(i)}, E_2^{(i)})$, $i = 1, 2, \ldots, r$, be lattice points in the integer lattice $\mathbb{Z}^2$ such that (3.1.1) and*

$$
2A^{(i)} + D \geq 0 \quad \text{and} \quad E_1^{(i)} \geq E_2^{(i)}, \quad i = 1, 2, \ldots, r,
$$

hold. Let γ be an integer which satisfies the inequalities

$$
D - E_1^{(1)} \leq \gamma \leq \min_{1 \leq i \leq r}(E_2^{(i)} + i - 1), \quad \max_{1 \leq i \leq r}(D - E_2^{(i)} - i) \leq \gamma \leq E_1^{(1)} - 1,
$$
$$
\tag{4.2.1a}
$$
$$
\text{and} \; -A^{(1)} \leq \gamma \leq A^{(1)} + D. \tag{4.2.1b}
$$

The generating function $\sum q^{\mathrm{ymaj}_{\beta;\gamma}}(\mathfrak{P})$, where the sum is over all nonintersecting families $\mathfrak{P} = (P_1, \ldots, P_r)$ of lattice paths which lie below the line $x = y$, $P_i : \mathcal{A}_i \to \mathcal{E}_i$, $i = 1, 2, \ldots, r$, and where $\mathrm{ymaj}_{\beta;\gamma}(\mathfrak{P})$ is given by (4.0.8), is equal to the expression

$$
\det_{1 \leq s,t \leq r} \left(q^{s(A^{(s)} - A^{(t)})} \left(\sum_{j \geq 0} q^{j(j+\beta+\gamma+A^{(t)} - s + 1)} \begin{bmatrix} -\beta \\ j \end{bmatrix} \begin{bmatrix} \beta + E_1^{(s)} + E_2^{(s)} - D \\ E_1^{(s)} - A^{(t)} - D - j \end{bmatrix} \right.\right.
$$
$$
\left.\left. - q^{s(2A^{(t)} + D + \beta + 1)} \sum_{j \geq 0} q^{j(j+\beta+\gamma+A^{(t)} + s + 1)} \begin{bmatrix} -\beta \\ j \end{bmatrix} \begin{bmatrix} \beta + E_1^{(s)} + E_2^{(s)} - D \\ E_2^{(s)} - A^{(t)} - D - j - 1 \end{bmatrix} \right)\right).
$$
$$
\tag{4.2.2}
$$

PROOF. We adopt the notations of subsection 2.2. Moreover we use the phrases "paths of reflected type" and "paths of ordinary type" which were introduced at the beginning of subsection 3.2. We follow the basic framework of the proof of Theorem 5. The details however are more delicate.

First we have to define a weight function on the set (2.2.14). We are going to use Proposition 24. Relation (5.2.2) tells us that if a path is of reflected type we have to take xmaj instead of ymaj. This observation motivates to introduce a weight function ϑ_3 by

$$\vartheta_3(\mathfrak{P}) = (-1)^{\|\eta\|} \operatorname{sgn} \sigma \, q^{\sum_{i=1}^{r}\left(i(A^{(i)}-A^{(\sigma(i))})+i\eta_i(2A^{(\sigma(i))}+D+\beta+1)\right)} q^{\operatorname{ymaj}_{\beta;\gamma}^{(\eta)}(\mathfrak{P})}$$
$$\text{if } \mathfrak{P} \in Q(\mathfrak{R}^\eta \mathbf{A}_\sigma \to \mathbf{E}), \quad (4.2.3)$$

where $\operatorname{ymaj}_{\beta;\gamma}^{(\eta)}$ is given by

$$\operatorname{ymaj}_{\beta;\gamma}^{(\eta)}(\mathfrak{P}) = \sum_{i=1}^{r}\left((1-\eta_i)\operatorname{ymaj}_{\beta;\gamma-i+1}(P_i) + \eta_i(\operatorname{xmaj}_{\beta;\gamma+i}(P_i))\right). \quad (4.2.4)$$

The effect of (4.2.4) is that P_i is counted by ymaj if P_i is of ordinary type and by xmaj if P_i is of reflected type, just as desired. Since for $\eta = \mathbf{0}$ all paths P_i are of ordinary type, and the i-th path is counted by $\operatorname{ymaj}_{\beta;\gamma-i+1}$, because of (4.0.8) the generating function in question is $GF(Q(\mathbf{A} \to \mathbf{E})^+; \vartheta_3)$. Obviously the equation

$$GF(Q(\mathbf{A} \to \mathbf{E})^+; \vartheta_3) = GF(Q(\mathbf{A} \to \mathbf{E}); \vartheta_3) - GF(Q(\mathbf{A} \to \mathbf{E})^-; \vartheta_3) \quad (4.2.5)$$

holds. This is the extension of (3.2.3).

We are going to construct an involution Θ_3 on the set in (2.2.13) which is sign-reversing with respect to ϑ_3. Let $\mathfrak{P} = (P_1, \ldots, P_r)$ be an element of $Q(\mathfrak{R}^\eta \mathbf{A}_\sigma \to \mathbf{E})^-$. Consider all meeting points of the path P_1 with the line $x = y - 1$ and all meeting points of neighbouring paths P_I, P_{I+1} with the restriction that if P_I and P_{I+1} are of different type (which means that neither both are of ordinary type nor both are of reflected type), then the paths $P_1, \ldots, P_{I-1}$ are pairwise nonintersecting and lie below $x = y$, and both P_I and P_{I+1}, after having met for the last time, do not meet one of the paths $P_1, \ldots, P_{I-1}$. In addition, if P_I is of ordinary type and P_{I+1} is of reflected type then $\sigma(I) \geq I$, and if P_I is of reflected type and P_{I+1} is of ordinary type then $\sigma(I + 1) \geq I$. Note that, compared to the proof of Theorem 5, here we refine the choice of meeting points between neighbouring paths. That there is at least one meeting point of the above kind is guaranteed by the strong statement of Lemma 4: Either there is a meeting point of P_1 with $x = y - 1$, or there is an I such that P_I and P_{I+1} intersect and such that P_I is of ordinary type. Should P_{I+1} be of reflected type then, also by Lemma 4, $P_1, \ldots, P_I$ are pairwise nonintersecting and lie below $x = y$. So in particular, P_I does not meet $P_1, \ldots, P_{I-1}$ at all. Besides, since due to (3.1.1b) P_I lies below $P_1, \ldots, P_{I-1}$, and the final point of P_{I+1} lies in the South-East of the final point of P_I, P_{I+1} must stay below P_I and hence below $P_1, \ldots, P_{I-1}$ after having met P_I for the last time. That $\sigma(I) \geq I$ holds is seen by observing that (3.1.1a) and the fact that P_2 lies below P_1, P_3 below P_2, ..., P_I below P_{I-1}, imply that $1 \le \sigma(1) < \sigma(2) < \cdots < \sigma(I)$.

Now choose the right-most and among them the highest meeting point and denote it by $\mathcal{S}$.

Again two cases have to be considered for the definition of Θ_3. If $\mathcal{S}$ is a meeting point between two neighbouring paths the definition of Θ_3 is identical with that of Θ_1 in (3.2.4a,b). Namely, among the pairs P_I, P_{I+1} which meet in $\mathcal{S}$ choose that one with I being minimal and set

$$\Theta_3(\mathfrak{P}) := (P_1, \ldots, P'_I, P'_{I+1}, \ldots, P_r) \,, \tag{4.2.6a}$$

where the paths P'_I, P'_{I+1} are given by

$$(P'_I, P'_{I+1}) = \begin{cases} \Psi((P_I, P_{I+1})) & \text{if } \sigma(I) < \sigma(I+1) \\ \Psi^{-1}((P_I, P_{I+1})) & \text{if } \sigma(I) > \sigma(I+1) \end{cases}. \tag{4.2.6b}$$

The map Ψ is defined in Proposition 27. Recall that $\mathcal{S}$ is the last meeting point between P_I and P_{I+1} and that with (4.2.6a,b) the family $\Theta_3(\mathfrak{P})$ is an element of $Q(\mathfrak{R}^{\eta^{(I,I+1)}} \mathbf{A}_{\sigma(I,I+1)} \to \mathbf{E})^-$, where $\eta^{(I,I+1)} = (\eta_1, \ldots, \eta_{I+1}, \eta_I, \ldots, \eta_r)$.

On the other hand, if $\mathcal{S}$ is no meeting point between neighbouring paths of the above type, then we define

$$\Theta_3(\mathfrak{P}) := (P''_1, P_2, \ldots, P_r), \tag{4.2.6c}$$

where the path P''_1 is given by

$$P''_1 = \begin{cases} \Phi^{(1)}_\gamma(P_1) & \text{if } \eta_1 = 0 \\ \left(\Phi^{(1)}_\gamma\right)^{-1}(P_1) & \text{if } \eta_1 = 1. \end{cases} \tag{4.2.6d}$$

The map $\Phi^{(1)}_\gamma$ is defined in Proposition 24. To justify that P''_1 is well-defined, note that because of (3.1.1a) and (4.2.1b) we have

$$-A^{(\sigma(I))} \le -A^{(1)} \le \gamma \le A^{(1)} + D \le A^{(\sigma(I))} + D,$$

which implies $-A^{(\sigma(I))} \le \gamma \le A^{(\sigma(I))} + D$. But $(A^{(\sigma(I))} + D, -A^{(\sigma(I))})$ is the starting point of P_I for $\eta_I = 0$ or, respectively, it is the reflected (reflected in the line $x = y - 1$) starting point of P_I for $\eta_I = 1$. This says nothing but that in both cases (5.2.1) is satisfied and so $\Phi^{(1)}_\gamma(P_1)$, respectively $(\Phi^{(1)}_\gamma)^{(-1)}(P_1)$ indeed are well-defined. Obviously with (4.2.6c,d) the family $\Theta_3(\mathfrak{P})$ is an element of $Q(\mathfrak{R}^{\eta^{(1)}} \mathbf{A}_\sigma \to \mathbf{E})^-$, where $\eta^{(1)} = (1 - \eta_1, \eta_2, \ldots, \eta_r)$.

It is not too difficult to see that Θ_3 is an involution. There are only two non-trivial cases. One of these two non-trivial cases arises if $\mathcal{S}$ is the last meeting point between P_I and P_{I+1}, I being minimal, where P_I is of ordinary type and P_{I+1} is of reflected type. In this case because of our restricted choice of meeting points the paths

$P_1, \ldots, P_{I-1}$ must be pairwise nonintersecting and lie below $x = y$. In addition, the portions of P_I and P_{I+1} after their last meeting point $\mathcal{S}$ do not meet $P_1, \ldots, P_{I-1}$. By (4.2.6a,b), P_I' is of reflected type and P_{I+1}' is of ordinary type. Since Ψ has the property of "changing nothing after $\mathcal{S}$" the paths P_I' and P_{I+1}', after having met in $\mathcal{S}$ for the last time, do not meet $P_1, \ldots, P_{I-1}$. Besides, $(\sigma(I, I+1))(I+1) = \sigma(I) \geq I$. Hence $\mathcal{S}$, being a meeting point between P_I' and P_{I+1}', is also under consideration for applying Θ_3 to $\Theta_3(\mathfrak{P}) = (P_1, \ldots, P_I', P_{I+1}', \ldots, P_r)$. But then it is the right-most and highest meeting point for $\Theta_3(\mathfrak{P})$. Therefore when applying Θ_3 to $\Theta_3(\mathfrak{P})$ one obtains $\mathfrak{P}$ again. The case of P_I being of reflected type and P_{I+1} being of ordinary type is dealt with similarly.

We want to establish that Θ_3 is sign-reversing with respect to ϑ_3. First assume that $\Theta_3(\mathfrak{P})$ is obtained using (4.2.6a,b). There are four cases depending on P_I being of ordinary or reflected type and on P_{I+1} being of ordinary or reflected type. If both P_I and P_{I+1} are of ordinary type, i.e. $\eta_I = \eta_{I+1} = 0$, from (5.3.2) we get

$$\mathrm{ymaj}_{\beta; \gamma - I + 1}(P_I) + \mathrm{ymaj}_{\beta; \gamma - I}(P_{I+1})$$
$$= \mathrm{ymaj}_{\beta; \gamma - I + 1}(P_I') + \mathrm{ymaj}_{\beta; \gamma - I}(P_{I+1}') + A^{(\sigma(I+1))} - A^{(\sigma(I))}$$

and hence

$$\mathrm{ymaj}_{\beta; \gamma}^{(\eta)}(\mathfrak{P}) = \mathrm{ymaj}_{\beta; \gamma}^{(\eta^{(I, I+1)})}(\Theta_3(\mathfrak{P})) + A^{(\sigma(I+1))} - A^{(\sigma(I))}$$
$$= \mathrm{ymaj}_{\beta; \gamma}^{(\eta^{(I, I+1)})}(\Theta_3(\mathfrak{P})) + A^{(\sigma(I+1))} - \eta_{I+1}(2A^{(\sigma(I+1))} + D + \beta + 1)$$
$$- A^{(\sigma(I))} + \eta_I(2A^{(\sigma(I))} + D + \beta + 1).$$

Secondly, if both P_I and P_{I+1} are of reflected type, i.e. $\eta_I = \eta_{I+1} = 1$, then from (5.3.3) we get

$$\mathrm{xmaj}_{\beta; \gamma + I}(P_I) + \mathrm{xmaj}_{\beta; \gamma + I + 1}(P_{I+1})$$
$$= \mathrm{xmaj}_{\beta; \gamma + I}(P_I') + \mathrm{xmaj}_{\beta; \gamma + I + 1}(P_{I+1}') - A^{(\sigma(I+1))} + A^{(\sigma(I))}$$

and hence

$$\mathrm{ymaj}_{\beta; \gamma}^{(\eta)}(\mathfrak{P}) = \mathrm{ymaj}_{\beta; \gamma}^{(\eta^{(I, I+1)})}(\Theta_3(\mathfrak{P})) - A^{(\sigma(I+1))} + A^{(\sigma(I))}$$
$$= \mathrm{ymaj}_{\beta; \gamma}^{(\eta^{(I, I+1)})}(\Theta_3(\mathfrak{P})) + A^{(\sigma(I+1))} - \eta_{I+1}(2A^{(\sigma(I+1))} + D + \beta + 1)$$
$$- A^{(\sigma(I))} + \eta_I(2A^{(\sigma(I))} + D + \beta + 1).$$

Now comes the first delicate case. Let P_I be of reflected and P_{I+1} be of ordinary type, i.e. $\eta_I = 1$ and $\eta_{I+1} = 0$. We want to apply (5.3.5) with $\gamma_1 = \gamma + I$ and $\gamma_2 = \gamma - I$. In order to be allowed to use (5.3.5), according to Proposition 27 it suffices to check that $A + D < \gamma + I$, that the last meeting point $\mathcal{S}$ of P_1 and P_2 is located to the right or on the vertical line $x = \gamma + I$, and that $-B - 1 \leq \gamma - I < -A$, where $(A + D, -A)$ is the starting point of P_I and $(B + D, -B)$ is the starting point of P_{I+1}. Recall that because of $\mathfrak{P} \in Q(\mathfrak{R}^{\eta} \mathbf{A}_{\sigma} \to \mathbf{E})^{-}$ the starting point of P_I is $(-A^{(\sigma(I))}-1, A^{(\sigma(I))}+D+1)$, while the starting point of P_{I+1} is $(A^{(\sigma(I+1))} + D, -A^{(\sigma(I+1))})$. Hence the first inequality trivially holds because of (4.2.1b), $-A^{(\sigma(I))} - 1 < -A^{(1)} \leq \gamma < \gamma + I$. To continue, let $\mathcal{S} = (S_1, S_2)$. Since $\mathcal{S}$ is on P_I we must have $S_2 \geq A^{(\sigma(I))} + D + 1$. It is easy to see that $\mathcal{S}$ must lie below or on $x = y$. From the assumption that $P_1, \ldots, P_{I-1}$ are pairwise nonintersecting and that the portions of P_I and P_{I+1} after $\mathcal{S}$ do not meet $P_1, \ldots, P_{I-1}$, we infer that P_1 passes through $y = S_2$ left-most, that P_2 passes $y = S_2$ to the right of P_1, $\ldots$, P_{I-1} to the right of P_{I-2}, that moreover $P_1, \ldots, P_{I-1}$ have no point in common and that $\mathcal{S}$ is located strictly to the right of the passage of P_{I-1} through $y = S_2$. Besides, because of lying below $x = y$, P_1 leaves $y = S_2$ earliest in $(S_2 + 1, S_2)$. Thus we infer $S_1 \geq S_2 + I$. These observations together with (3.1.1a) and (4.2.1b) yield

$$\gamma + I \leq A^{(1)} + D + I < A^{(\sigma(I))} + D + I + 1 \leq S_2 + I \leq S_1,$$

which says nothing but that $\mathcal{S}$ is located to the right of $x = \gamma + I$. Besides, because of $\sigma(I+1) \geq I$ (this is due to the assumption of P_I and P_{I+1} being of different type; cf. the paragraph after (4.2.5)), (3.1.1a), and (4.2.1b) we have

$$-A^{(\sigma(I+1))} - 1 \leq -A^{(I)} - 1 \leq -A^{(1)} - I \leq \gamma - I \leq \gamma \leq A^{(1)} + D < A^{(\sigma(I))} + D + 1,$$

which in particular implies $-A^{(\sigma(I+1))} - 1 \leq \gamma - I \leq A^{(\sigma(I))} + D + 1$, which is just the second inequality required to hold in order to be allowed to apply (5.3.5). Hence we may indeed apply (5.3.5) to get

$$\begin{aligned}
&\text{xmaj}_{\beta;\gamma+I}(P_I) + \text{ymaj}_{\beta;\gamma-I}(P_{I+1}) \\
&\quad = \text{ymaj}_{\beta;\gamma-I+1}(P_I') + \text{xmaj}_{\beta;\gamma+I+1}(P_{I+1}') + A^{(\sigma(I+1))} + A^{(\sigma(I))} + D + 1 + \beta,
\end{aligned}$$

and hence

$$\begin{aligned}
\text{ymaj}_{\beta;\gamma}^{(\eta)}(\mathfrak{P}) &= \text{ymaj}_{\beta;\gamma}^{(\eta^{(I,I+1)})}(\Theta_3(\mathfrak{P})) + A^{(\sigma(I+1))} + A^{(\sigma(I))} + D + 1 + \beta \\
&= \text{ymaj}_{\beta;\gamma}^{(\eta^{(I,I+1)})}(\Theta_3(\mathfrak{P})) + A^{(\sigma(I+1))} - \eta_{I+1}(2A^{(\sigma(I+1))} + D + \beta + 1) \\
&\qquad - A^{(\sigma(I))} + \eta_I(2A^{(\sigma(I))} + D + \beta + 1).
\end{aligned}$$

Finally, let P_I be of ordinary type and P_{I+1} be of reflected type, i.e. $\eta_I = 0$ and $\eta_{I+1} = 1$. Again we want to apply (5.3.5) with $\gamma_1 = \gamma + I$ and $\gamma_2 = \gamma - I$, but this time in the reverse direction, meaning that we choose $(Q_1, Q_2) = (P_I, P_{I+1})$. The starting point of P_I now is $(A^{(\sigma(I))}+D, -A^{(\sigma(I))})$, and that of P_{I+1} is $(-A^{(\sigma(I+1))}-1, A^{(\sigma(I+1))}+D+1)$. Therefore in Proposition 27 we have to choose $A = -A^{(\sigma(I+1))}-D-1$ and $B = A^{(\sigma(I))}$. The first inequality required for (5.3.5) is easily shown by again using (3.1.1a) and (4.2.1b), $-A^{(\sigma(I+1))} - 1 < -A^{(1)} \leq \gamma \leq \gamma + I$. Now, as before let $\mathcal{S} = (S_1, S_2)$. Since $\mathcal{S}$ is on P_{I+1} we must have $S_2 \geq A^{(\sigma(I+1))} + D + 1$. In the same manner as before we conclude $S_1 \geq S_2 + I$. Combining these observations with (3.1.1a) and (4.2.1b) gives

$$\gamma + I \leq A^{(1)} + D + I < A^{(\sigma(I+1))} + D + I + 1 \leq S_2 + I \leq S_1,$$

which implies that $\mathcal{S}$ is located to the right of $x = \gamma + I$. Besides, because of $\sigma(I) \geq I$ (cf. the paragraph after (4.2.5)), (3.1.1a), and (4.2.1b) we have

$$-A^{(\sigma(I))} - 1 \leq -A^{(I)} - 1 \leq -A^{(1)} - I \leq \gamma - I \leq \gamma \leq A^{(1)} + D < A^{(\sigma(I+1))} + D + 1,$$

which in particular implies $-A^{(\sigma(I))} - 1 \leq \gamma - I \leq A^{(\sigma(I+1))} + D + 1$, which is the second inequality required for (5.3.5). Hence we may apply (5.3.5) to obtain

$$\begin{aligned}
\mathrm{xmaj}_{\beta;\gamma+I}&(P'_I) + \mathrm{ymaj}_{\beta;\gamma-I}(P'_{I+1}) \\
&= \mathrm{ymaj}_{\beta;\gamma-I+1}(P_I) + \mathrm{xmaj}_{\beta;\gamma+I+1}(P_{I+1}) + A^{(\sigma(I))} + A^{(\sigma(I+1))} + D + 1 + \beta,
\end{aligned}$$

and hence

$$\begin{aligned}
\mathrm{ymaj}_{\beta;\gamma}^{;(\eta^{(I,I+1)})}(\Theta_3(\mathfrak{P})) &= \mathrm{ymaj}_{\beta;\gamma}^{(\eta)}(\mathfrak{P}) + A^{(\sigma(I))} + A^{(\sigma(I+1))} + D + 1 + \beta \\
&= \mathrm{ymaj}_{\beta;\gamma}^{(\eta)}(\mathfrak{P}) - A^{(\sigma(I+1))} + \eta_{I+1}(2A^{(\sigma(I+1))} + D + \beta + 1) \\
&\quad + A^{(\sigma(I))} - \eta_I(2A^{(\sigma(I))} + D + \beta + 1).
\end{aligned}$$

Summarizing, in all four cases we have

$$\begin{aligned}
\mathrm{ymaj}_{\beta;\gamma}^{;(\eta^{(I,I+1)})}(\Theta_3(\mathfrak{P})) &= \mathrm{ymaj}_{\beta;\gamma}^{(\eta)}(\mathfrak{P}) + A^{(\sigma(I))} - \eta_I(2A^{(\sigma(I))} + D + \beta + 1) \\
&\quad - A^{(\sigma(I+1))} + \eta_{I+1}(2A^{(\sigma(I+1))} + D + \beta + 1).
\end{aligned}$$
$$(4.2.7)$$

Now assume that $\Theta_3(\mathfrak{P})$ is obtained by (4.2.6a,b). Using (4.2.7) we modify (3.2.6) to get

$$\begin{aligned}
\vartheta_3(\Theta_3(\mathfrak{P})) &= (-1)^{\|\eta^{(I,I+1)}\|} \, \mathrm{sgn}\left(\sigma(I, I+1)\right) \\
&\quad \cdot q^{\sum_{i=1}^r \left(i(A^{(i)} - A^{((\sigma(I,I+1))(i))}) + i\eta_i^{(I,I+1)}(2A^{((\sigma(I,I+1))(i))} + D + \beta + 1)\right)} q^{\mathrm{ymaj}_{\beta;\gamma}^{(\eta^{(I,I+1)})}} \Theta_3(\mathfrak{P}) \\
&= -(-1)^{\|\eta\|} \, \mathrm{sgn}\,\sigma \, q^{\sum_{i=1}^r i(A^{(i)} - A^{(\sigma(i))}) + I(A^{(\sigma(I))} - A^{(\sigma(I+1))}) + (I+1)(A^{(\sigma(I+1))} - A^{(\sigma(I))})} \\
&\quad \cdot q^{\sum_{i=1}^r i\eta_i(2A^{(\sigma(i))} + D + \beta + 1) - \eta_{I+1}(2A^{(\sigma(I+1))} + D + \beta + 1) + \eta_I(2A^{(\sigma(I))} + D + \beta + 1)} \\
&\quad \cdot q^{\mathrm{ymaj}_{\beta;\gamma}^{(\eta)}} \mathfrak{P} + \left(A^{(\sigma(I))} - \eta_I(2A^{(\sigma(I))} + D + \beta + 1)\right) - \left(A^{(\sigma(I+1))} - \eta_{I+1}(2A^{(\sigma(I+1))} + D + \beta + 1)\right) \\
&= -\vartheta_3(\mathfrak{P}).
\end{aligned}$$
$$(4.2.8)$$

Next we assume that $\Theta_3(\mathfrak{P})$ is obtained using (4.2.6c,d). If $\eta_1 = 0$, using (3.2.5) and (5.2.2), we get

$$\mathrm{xmaj}_{\beta;\gamma+1}(\Phi_\gamma^{(1)}(P_1)) = \mathrm{ymaj}_{\beta;\gamma}(P_1) - (A^{(\sigma(1))} + D) - A^{(\sigma(1))} - 1 - \beta.$$

Consequently, by the definitions of ϑ_3 (cf. (4.2.3)) and of $\mathrm{ymaj}_{\beta;\gamma}^{(\eta)}$ (cf. (4.2.4)) we obtain

$$\vartheta_3(\Theta_3(\mathfrak{P}))$$

$$= (-1)^{\|\eta^{(1)}\|} \operatorname{sgn} \sigma \, q^{\sum_{i=1}^{r} \left(i(A^{(i)} - A^{(\sigma(i))}) + i\eta_i^{(1)}(2A^{(\sigma(i))} + D + \beta + 1) \right)} q^{\mathrm{ymaj}_{\beta;\gamma}^{(\eta^{(1)})}} \Theta_3(\mathfrak{P})$$

$$= -(-1)^{\|\eta\|} \operatorname{sgn} \sigma \, q^{\sum_{i=1}^{r} i(A^{(i)} - A^{(\sigma(i))})} q^{\sum_{i=1}^{r} i\eta_i(2A^{(\sigma(i))} + D + \beta + 1) + (2A^{(\sigma(1))} + D + \beta + 1)}$$

$$\cdot q^{\mathrm{ymaj}_{\beta;\gamma}^{(\eta)}} \mathfrak{P} - \left(A^{(\sigma(1))} + D \right) - A^{(\sigma(1))} - 1 - \beta$$

$$= -\vartheta_3(\mathfrak{P}). \tag{4.2.9}$$

Likewise, if $\eta_1 = 1$ by (3.2.5) and (5.2.2) we have

$$\mathrm{ymaj}_{\beta;\gamma}((\Phi_\gamma^{(1)})^{(-1)}(P_1)) = \mathrm{xmaj}_{\beta;\gamma+1}(P_1) + A^{(\sigma(1))} + D + A^{(\sigma(1))} + 1 + \beta.$$

Consequently, by the definitions (4.2.3) and (4.2.4) we obtain

$$\vartheta_3(\Theta_3(\mathfrak{P}))$$

$$= (-1)^{\|\eta^{(1)}\|} \operatorname{sgn} \sigma \, q^{\sum_{i=1}^{r} \left(i(A^{(i)} - A^{(\sigma(i))}) + i\eta_i^{(1)}(2A^{(\sigma(i))} + D + \beta + 1) \right)} q^{\mathrm{ymaj}_{\beta;\gamma}^{(\eta^{(1)})}} \Theta_3(\mathfrak{P})$$

$$= -(-1)^{\|\eta\|} \operatorname{sgn} \sigma \, q^{\sum_{i=1}^{r} i(A^{(i)} - A^{(\sigma(i))})} q^{\sum_{i=1}^{r} i\eta_i(2A^{(\sigma(i))} + D + \beta + 1) - (2A^{(\sigma(1))} + D + \beta + 1)}$$

$$\cdot q^{\mathrm{ymaj}_{\beta;\gamma}^{(\eta)}} \mathfrak{P} + A^{(\sigma(1))} + D + A^{(\sigma(1))} + 1 + \beta$$

$$= -\vartheta_3(\mathfrak{P}). \tag{4.2.10}$$

Now, since by (4.2.8), (4.2.9), (4.2.10) we finally settled that Θ_3 is sign-reversing with respect to ϑ_3, we may infer

$$\sum_{\sigma \in \mathfrak{S}_r, \, \eta \in \{0,1\}^r} GF(Q(\mathfrak{R}^\eta \mathbf{A}_\sigma \to \mathbf{E})^-; \vartheta_3) = 0 \,. \tag{4.2.11}$$

All that remains, is to analogize the computation (3.2.9). Indeed, by consecutive use of (4.2.5), (4.2.11), (2.2.16), (5.1.11) and (5.1.12),

$$GF(Q(\mathbf{A} \to \mathbf{E})^+; \vartheta_3) = GF(Q(\mathbf{A} \to \mathbf{E}); \vartheta_3) - GF(Q(\mathbf{A} \to \mathbf{E})^-; \vartheta_3)$$

$$= GF(Q(\mathbf{A} \to \mathbf{E}); \vartheta_3) + \sum_{\substack{\sigma \in \mathfrak{S}_r,\, \eta \in \{0,1\}^r \\ (\sigma,\eta) \neq (id,\mathbf{0})}} GF(Q(\mathfrak{R}^\eta \mathbf{A}_\sigma \to \mathbf{E})^-; \vartheta_3)$$

$$= \sum_{\sigma \in \mathfrak{S}_r,\, \eta \in \{0,1\}^r} GF(Q(\mathfrak{R}^\eta \mathbf{A}_\sigma \to \mathbf{E}); \vartheta_3)$$

$$= \sum_{\sigma \in \mathfrak{S}_r} \operatorname{sgn} \sigma \prod_{i=1}^r q^{i(A^{(i)} - A^{(\sigma(i))})} \big(GF(P(\mathcal{A}_{\sigma(i)} \to \mathcal{E}_i); \operatorname{ymaj}_{\beta; \gamma - i + 1})$$

$$- q^{i(2A^{(\sigma(i))} + D + \beta + 1)} GF(P(\mathfrak{R}\mathcal{A}_{\sigma(i)} \to \mathcal{E}_i); \operatorname{xmaj}_{\beta; \gamma + i}) \big)$$

$$= \sum_{\sigma \in \mathfrak{S}_r} \operatorname{sgn} \sigma \prod_{i=1}^r q^{i(A^{(i)} - A^{(\sigma(i))})}$$

$$\times \left(\sum_{j \geq 0} q^{j(j + \beta + \gamma - i + 1 + A^{(\sigma(i))})} \begin{bmatrix} -\beta \\ j \end{bmatrix} \begin{bmatrix} \beta + E_1^{(i)} + E_2^{(i)} - D \\ E_1^{(i)} - A^{(\sigma(i))} - D - j \end{bmatrix} \right.$$

$$- q^{i(2A^{(\sigma(i))} + D + \beta + 1)} \sum_{j \geq 0} q^{j(j + \beta + \gamma + i + A^{(\sigma(i))} + 1)}$$

$$\left. \begin{bmatrix} -\beta \\ j \end{bmatrix} \begin{bmatrix} \beta + E_1^{(i)} + E_2^{(i)} - D \\ E_2^{(i)} - A^{(\sigma(i))} - D - j - 1 \end{bmatrix} \right)$$

$$= \det_{1 \leq s, t \leq r} \left(q^{s(A^{(s)} - A^{(t)})} \left(\sum_{j \geq 0} q^{j(j + \beta + \gamma + A^{(t)} - s + 1)} \begin{bmatrix} -\beta \\ j \end{bmatrix} \begin{bmatrix} \beta + E_1^{(s)} + E_2^{(s)} - D \\ E_1^{(s)} - A^{(t)} - D - j \end{bmatrix} \right. \right.$$

$$- q^{s(2A^{(t)} + D + \beta + 1)} \sum_{j \geq 0} q^{j(j + \beta + \gamma + A^{(t)} + s + 1)}$$

$$\left. \left. \begin{bmatrix} -\beta \\ j \end{bmatrix} \begin{bmatrix} \beta + E_1^{(s)} + E_2^{(s)} - D \\ E_2^{(s)} - A^{(t)} - D - j - 1 \end{bmatrix} \right) \right).$$

Note that because of the restrictions (4.2.1a), the requirements being necessary for being allowed to use (5.1.11) and (5.1.12) are always satisfied. The last line is identical with (4.2.2).   □

If the paths lie above the line $x = y$, there is a result for the xmaj.

Theorem 17. Let $\mathcal{A}_i = (-A^{(i)}, A^{(i)} + D)$ and $\mathcal{E}_i = (E_1^{(i)}, E_2^{(i)})$, $i = 1, 2, \ldots, r$, be lattice points in the integer lattice $\mathbb{Z}^2$ such that (3.2.10) holds. Let γ be an integer *satisfying*

$$\max_{1 \le i \le r}(D - E_2^{(i)} + i - 1) \le \gamma \le \min_{1 \le i \le r}(E_1^{(i)} + i - 1),$$

$$\max_{1 \le i \le r}(D - E_1^{(i)} - i + 2) \le \gamma \le \min_{1 \le i \le r}(E_2^{(i)} - i + 2),$$

$$\text{and} \quad -A^{(1)} \le \gamma \le A^{(1)} + D. \quad (4.2.12)$$

The generating function $\sum q^{\text{xmaj}_{\beta;\gamma}(\mathfrak{P})}$, *where the sum is over all nonintersecting families* $\mathfrak{P} = (P_1, \ldots, P_r)$ *of lattice paths which lie above the line* $x = y$, $P_i : \mathcal{A}_i \to \mathcal{E}_i$, $i = 1, 2, \ldots, r$, *and where* $\text{xmaj}_{\beta;\gamma} \mathfrak{P}$ *is given by (4.0.9) (see the Remark below), is equal to the expression*

$$\det_{1 \le s,t \le r} \left(q^{s(A^{(s)} - A^{(t)})} \left(\sum_{j \ge 0} q^{j(j+\beta+\gamma+A^{(t)}-s+1)} \begin{bmatrix} -\beta \\ j \end{bmatrix} \begin{bmatrix} \beta + E_1^{(s)} + E_2^{(s)} - D \\ E_2^{(s)} - A^{(t)} - D - j \end{bmatrix} \right. \right.$$
$$\left. \left. - q^{(s-1)(2A^{(t)}+D+\beta+1)} \sum_{j \ge 0} q^{j(j+\beta+\gamma+A^{(t)}+s-1)} \begin{bmatrix} -\beta \\ j \end{bmatrix} \begin{bmatrix} \beta + E_1^{(s)} + E_2^{(s)} - D \\ E_1^{(s)} - A^{(t)} - D - j - 1 \end{bmatrix} \right) \right).$$

$$(4.2.13)$$

REMARK. Because of (3.2.10a,b) the paths $P_1, \ldots, P_r$ are ordered "from bottom to top". It is important to observe that since we made the definition (4.0.9) for $\text{xmaj}_{\beta;\gamma}$ of a family of paths, by requiring (2.2.2), dependent on the order of the paths, here in fact we have

$$\text{xmaj}_{\beta;\gamma}(\mathfrak{P}) = \sum_{i=1}^{r} \text{xmaj}_{\beta;\gamma-i+1}(P_i). \quad (4.2.14)$$

SKETCH OF PROOF. We follow the line of the proof of Theorem 16. The main difference is that instead of $\Phi_\gamma^{(1)}$ we have to use $\Phi^{(2)}$, which is defined in Proposition 26. Of course, everywhere the roles of ymaj and xmaj have to be exchanged. We only remark that the weight function to be used is

$$\vartheta_4(\mathfrak{P}) = (-1)^{\|\eta\|} \operatorname{sgn} \sigma \, q^{\sum_{i=1}^{r}\left(i(A^{(i)} - A^{(\sigma(i))}) + i\eta_i(2A^{(\sigma(i))} + D + \beta + 1)\right)} q^{\text{xmaj}_{\beta;\gamma}^{(\eta)}} \mathfrak{P}$$

$$\text{if } \mathfrak{P} \in Q(\mathfrak{R}^\eta \mathbf{A}_\sigma \to \mathbf{E}), \quad (4.2.15)$$

where $\text{xmaj}_{\beta;\gamma}^{(\eta)}$ is given by

$$\text{xmaj}_{\beta;\gamma}^{(\eta)}(\mathfrak{P}) = \sum_{i=1}^{r} \left((1 - \eta_i) \text{xmaj}_{\beta;\gamma-i+1}(P_i) + \eta_i(\text{ymaj}_{\beta;\gamma+i-2}(P_i)) \right). \quad (4.2.16)$$

What has to be determined, is the generating function $GF(\bar{Q}(\mathbf{A} \to \mathbf{E})^+, \vartheta_4)$, where $\bar{Q}(\mathbf{A} \to \mathbf{E})^+$ is the set of all families $\mathfrak{P} = (P_1, \ldots, P_r)$ of nonintersecting lattice paths which lie *above* $x = y$, $P_i : \mathcal{A}_i \to \mathcal{E}_i$. This is done quite analogously as in the proof of Theorem 16 and is therefore left to the reader.   □

Next we consider cases in which (4.2.2) or (4.2.13) can be simplified. With the special final points of Theorem 7, we can get rid of the determinant in (4.2.2) thus obtaining the multiple sum (4.2.17) and for the more general case the multiple sum (4.2.19). If $\beta = 1$ and $\gamma = (D-1)/2$, with the help of Proposition 36 the sum (4.2.17) can even be evaluated to result in a nice product. Since γ and D are integers, the relation $\gamma = (D-1)/2$ implies that in this case D has to be odd. There do not seem to be other special cases where (4.2.17) or (4.2.19) can be evaluated. In particular, for $r = 1$ none of the known basic hypergeometric summation theorems can be applied to (4.2.17) except for $\beta = 1$ and $\gamma = (D-1)/2$. So it is very unlikely that (4.2.17) can be summed for $r > 1$ in other cases.

Theorem 18. *Let* $\mathcal{A}_i = (A^{(i)} + D, -A^{(i)})$ *and* $\mathcal{E}_i = (E+i, E+2-i)$, $i = 1, 2, \ldots, r$, *be lattice points in the integer lattice* $\mathbb{Z}^2$ *such that (3.1.1a) and*

$$2A^{(i)} + D \geq 0, \quad i = 1, 2, \ldots, r,$$

hold. If γ *is an integer satisfying*

$$D - E - 1 \leq \gamma \leq E \quad \text{and} \quad -A^{(1)} \leq \gamma \leq A^{(1)} + D,$$

then the generating function $\sum q^{\mathrm{ymaj}_{\beta;\gamma}(\mathfrak{P})}$ *where the sum is over all nonintersecting families* $\mathfrak{P} = (P_1, \ldots, P_r)$ *of lattice paths which lie below the line* $x = y$, $P_i : \mathcal{A}_i \to \mathcal{E}_i$, $i = 1, 2, \ldots, r$, *is equal to the expression*

$$\sum_{k_1, \ldots, k_r \geq 0} \prod_{i=1}^{r} q^{k_i(k_i + \beta + \gamma + A^{(i)} - i + 1)} \begin{bmatrix} -\beta \\ k_i \end{bmatrix}$$

$$\frac{[\beta + 2E + 2i - D]!}{[E + r - A^{(i)} - k_i - D]! \, [\beta + E + r + 1 + A^{(i)} + k_i]!}$$

$$\prod_{1 \leq i < j \leq r} [A^{(j)} + k_j - A^{(i)} - k_i] \prod_{1 \leq i \leq j \leq r} [A^{(i)} + k_i + A^{(j)} + k_j + D + \beta + 1] . \quad (4.2.17)$$

In particular, if D is odd, say $D = 2D' + 1$, and if $-A^{(1)} \leq D' \leq E$, the generating function $\sum q^{\mathrm{ymaj}_{1;D'}(\mathfrak{P})}$ over the same set of nonintersecting families $\mathfrak{P}$ is given by

$$\prod_{i=1}^{r} \frac{[2E + 2i - 2D']!}{[E + r - A^{(i)} - 2D' - 1]! \, [E + r + 1 + A^{(i)}]!} \prod_{1 \leq i < j \leq r} [A^{(j)} - A^{(i)}]$$

$$\times \prod_{1 \leq i < j \leq r} [A^{(i)} + A^{(j)} + 2D' + 2] \prod_{i=1}^{r} \frac{[A^{(i)} + D' + 1]}{[E + i - D']} . \quad (4.2.18)$$

More generally, if D is again generic and if the final points of the lattice paths are $\mathcal{E}_i = (E_1 + i, E_2 - i)$, $i = 1, 2, \ldots, r$, with $E_1 \geq E_2 - 2$, then the generating function

$\sum q^{\mathrm{ymaj}_{\beta,\gamma}(\mathfrak{P})}$ for all nonintersecting families $\mathfrak{P} = (P_1, \ldots, P_r)$ of lattice paths which lie below the line $x = y$, $P_i : \mathcal{A}_i \to \mathcal{E}_i$, $i = 1, 2, \ldots, r$, is equal to the expression

$$\sum_{\eta \in \{0,1\}^r} (-1)^{\|\eta\|} \sum_{k_1,\ldots,k_r \geq 0} \prod_{i=1}^{r} \left(q^{i\eta_i(2A^{(i)}+2k_i+D+\beta+1)+k_i(k_i+\beta+\gamma+A^{(i)}-i+1)} \begin{bmatrix} -\beta \\ k_i \end{bmatrix} \right.$$
$$\left. \frac{[\beta + E_1 + E_2 + i - D - 1]!}{[\beta + E_2 + A^{(i)} + k_i - \eta_i(2A^{(i)} + 2k_i + D + \beta + 1) - 1]!} \right)$$
$$\frac{\displaystyle\prod_{1 \leq i < j \leq r} [A^{(j)} + k_j - \eta_j(2A^{(j)} + 2k_j + D + \beta + 1) - A^{(i)} - k_i + \eta_i(2A^{(i)} + 2k_i + D + \beta + 1)]}{\displaystyle\prod_{i=1}^{r} [E_1 + r - D - A^{(i)} - k_i + \eta_i(2A^{(i)} + 2k_i + D + \beta + 1)]!}$$

$$(4.2.19)$$

PROOF. In order to confirm (4.2.17), in Theorem 16 set $E_1^{(i)} = E + i$ and $E_2^{(i)} = E + 2 - i$ to obtain for the desired generating function the expression

$$\det \left(q^{s(A^{(s)} - A^{(t)})} \left(\sum_{j \geq 0} q^{j(j+\beta+\gamma+A^{(t)}-s+1)} \begin{bmatrix} -\beta \\ j \end{bmatrix} \begin{bmatrix} \beta + 2E + 2 - D \\ E + s - A^{(t)} - D - j \end{bmatrix} \right. \right.$$
$$\left. \left. - q^{s(2A^{(t)}+D+\beta+1)} \sum_{j \geq 0} q^{j(j+\beta+\gamma+A^{(t)}+s+1)} \begin{bmatrix} -\beta \\ j \end{bmatrix} \begin{bmatrix} \beta + 2E + 2 - D \\ E - s - A^{(t)} - D - j + 1 \end{bmatrix} \right) \right).$$

As before in the proof of Theorem 15, we denote the summation indices in the t-th column by k_t and then use the linearity in the t-th column to obtain that the last expression equals

$$\sum_{k_1,\ldots,k_r \geq 0} \prod_{i=1}^{r} q^{k_i(k_i+\beta+\gamma+A^{(i)}-i+1)} \begin{bmatrix} -\beta \\ k_i \end{bmatrix}$$
$$\det \left(q^{s(A^{(s)}+k_s-A^{(t)}-k_t)} \left(\begin{bmatrix} \beta + 2E + 2 - D \\ E + s - A^{(t)} - k_t - D \end{bmatrix} \right. \right.$$
$$\left. \left. - q^{s(2A^{(t)}+2k_t+D+\beta+1)} \begin{bmatrix} \beta + 2E + 2 - D \\ E - s - A^{(t)} - k_t - D + 1 \end{bmatrix} \right) \right).$$

The determinant in this last expression has already been evaluated. This is seen by replacing E by $E + \beta$, D by $D + \beta$, $A^{(t)}$ by $A^{(t)} + k_t$ in the proof of Theorem 7. The result is the expression (3.2.14), with these replacements. This immediately settles (4.2.17).

In order to establish (4.2.18), in (4.2.17) set $\beta = 1$, $D = 2D' + 1$, $\gamma = D'$. Use hypergeometric notation to obtain for the desired generating function the expression

$$\prod_{i=1}^{r} \frac{[2E + 2i - 2D']!}{[E + r - A^{(i)} - 2D' - 1]! \, [E + r + 2 + A^{(i)}]!}$$

$$\times \sum_{k_1, \ldots, k_r \geq 0} \prod_{i=1}^{r} q^{k_i(1 + E - D' + r - i)} \frac{(q^{A^{(i)} + 2D' + 1 - E - r})_{k_i}}{(q^{E + r + 3 + A^{(i)}})_{k_i}}$$

$$\prod_{1 \leq i < j \leq r} (1 - q^{A^{(j)} + k_j - A^{(i)} - k_i}) \prod_{1 \leq i \leq j \leq r} (1 - q^{A^{(i)} + k_i + A^{(j)} + k_j + 2D' + 3}) . \quad (4.2.20)$$

By setting $m_i = q^{A^{(i)} + D' + 3/2}$, $i = 1, 2, \ldots, r$, and $A = q^{D' - E - r - 1/2}$, this can be written in the form

$$\prod_{i=1}^{r} \frac{[2E + 2i - 2D']!}{[E + r - A^{(i)} - 2D' - 1]! \, [E + r + 2 + A^{(i)}]!}$$

$$\times \prod_{1 \leq i < j \leq r} [A^{(j)} - A^{(i)}] \prod_{1 \leq i \leq j \leq r} [A^{(i)} + A^{(j)} + 2D' + 3]$$

$$\times \sum_{k_1, \ldots, k_r \geq 0} \prod_{i=1}^{r} \left(\frac{\sqrt{q}}{q^i A} \right)^{k_i} \prod_{i=1}^{r} \frac{(m_i A)_{k_i}}{(q m_i / A)_{k_i}}$$

$$\times \prod_{1 \leq i < j \leq r} \frac{1 - \frac{m_j}{m_i} q^{k_j - k_i}}{1 - \frac{m_j}{m_i}} \prod_{1 \leq i \leq j \leq r} \frac{1 - m_i m_j q^{k_i + k_j}}{1 - m_i m_j} .$$

The r-fold sum is evaluated using (5.6.1). Thus we obtain

$$\prod_{i=1}^{r} \frac{[2E + 2i - 2D']!}{[E + r - A^{(i)} - 2D' - 1]! \, [E + r + 2 + A^{(i)}]!}$$

$$\times \prod_{1 \leq i < j \leq r} [A^{(j)} - A^{(i)}] \prod_{1 \leq i \leq j \leq r} [A^{(i)} + A^{(j)} + 2D' + 3]$$

$$\times \prod_{1 \leq i < j \leq r} \frac{[A^{(i)} + A^{(j)} + 2D' + 2]}{[A^{(i)} + A^{(j)} + 2D' + 3]} \prod_{i=1}^{r} \left(\frac{[A^{(i)} + D' + 1]}{[2A^{(i)} + 2D' + 3]} \frac{[E + A^{(i)} + r + 2]}{[E + r - D' - i + 1]} \right),$$

which simplifies to (4.2.18).

The proof of (4.2.19) proceeds in analogy with the proof of (3.2.15) and is therefore omitted.   □

The analogue of Theorem 18 which corresponds to Theorem 17 is the following result. A closed form is obtained for $E_1^{(i)} = E + 2 - i$, $E_2^{(i)} = E + i$, $\beta = 1$ and $\gamma = (D + 1)/2$, which forces D to be odd here, too. Again, obtaining a closed form for other choices of the parameters seems to be very unlikely.

Theorem 19. *Let $\mathcal{A}_i = (-A^{(i)}, A^{(i)}+D)$ and $\mathcal{E}_i = (E+2-i, E+i)$, $i = 1, 2, \ldots, r$, be lattice points in the integer lattice $\mathbb{Z}^2$ such that (3.2.10a) and*

$$2A^{(i)} + D \geq 0, \quad i = 1, 2, \ldots, r,$$

hold. If γ is an integer satisfying

$$D - E \leq \gamma \leq E + 1 \quad \text{and} \quad -A^{(1)} \leq \gamma \leq A^{(1)} + D,$$

then the generating function $\sum q^{\mathrm{xmaj}_{\beta;\gamma}}(\mathfrak{P})$ where the sum is over all nonintersecting families $\mathfrak{P} = (P_1, \ldots, P_r)$ of lattice paths which lie above the line $x = y$, $P_i : \mathcal{A}_i \to \mathcal{E}_i$, $i = 1, 2, \ldots, r$ is equal to the expression

$$q^{\binom{r+1}{2}+r(E-D)-\sum_{i=1}^{r} A^{(i)}} \sum_{k_1,\ldots,k_r \geq 0} \prod_{i=1}^{r} q^{k_i(k_i+\beta+\gamma+A^{(i)}-i)}$$

$$\begin{bmatrix} -\beta \\ k_i \end{bmatrix} \frac{[\beta + 2E + 2i - D]!}{[E + r - A^{(i)} - k_i - D]! \, [\beta + E + r + 1 + A^{(i)} + k_i]!}$$

$$\prod_{1 \leq i < j \leq r} [A^{(j)} + k_j - A^{(i)} - k_i] \prod_{1 \leq i < j \leq r} [A^{(i)} + k_i + A^{(j)} + k_j + D + \beta + 1]. \quad (4.2.21)$$

In particular, if D is odd, say $D = 2D' + 1$, and if $-A^{(1)} \leq D' \leq E$, the generating function $\sum q^{\mathrm{xmaj}_{1;D'+1}}(\mathfrak{P})$ over the same set of nonintersecting families $\mathfrak{P}$ is given by

$$q^{\binom{r+1}{2}+r(E-D)-\sum_{i=1}^{r} A^{(i)}} \prod_{i=1}^{r} \frac{[2E + 2i - 2D']!}{[E + r - A^{(i)} - 2D' - 1]! \, [E + r + 1 + A^{(i)}]!}$$

$$\times \prod_{1 \leq i < j \leq r} [A^{(j)} - A^{(i)}] \prod_{1 \leq i < j \leq r} [A^{(i)} + A^{(j)} + 2D' + 2] \prod_{i=1}^{r} \frac{[A^{(i)} + D' + 1]}{[E + i - D']}. \quad (4.2.22)$$

More generally, if D is again generic and if the final points of the lattice paths are $\mathcal{E}_i = (E_1 - i, E_2 + i)$, $i = 1, 2, \ldots, r$, with $E_1 \leq E_2 + 2$, then the generating function $\sum q^{\mathrm{xmaj}_{\beta;\gamma}}(\mathfrak{P})$ for all nonintersecting families $\mathfrak{P} = (P_1, \ldots, P_r)$ of lattice paths which lie above the line $x = y$, $P_i : \mathcal{A}_i \to \mathcal{E}_i$, $i = 1, 2, \ldots, r$, is equal to the expression

$$\sum_{\eta \in \{0,1\}^r} (-1)^{\|\eta\|} \sum_{k_1,\ldots,k_r \geq 0} \prod_{i=1}^{r} \left(q^{(i-1)\eta_i(2A^{(i)}+2k_i+D+\beta+1)+k_i(k_i+\beta+\gamma+A^{(i)}-i+1)} \begin{bmatrix} -\beta \\ k_i \end{bmatrix} \right.$$

$$\left. \frac{[\beta + E_1 + E_2 + i - D - 1]!}{[\beta + E_1 + A^{(i)} + k_i - \eta_i(2A^{(i)} + 2k_i + D + \beta + 1) - 1]!} \right)$$

$$\frac{\prod_{1 \leq i < j \leq r} [A^{(j)} + k_j - \eta_j(2A^{(j)} + 2k_j + D + \beta + 1) - A^{(i)} - k_i + \eta_i(2A^{(i)} + 2k_i + D + \beta + 1)]}{\prod_{i=1}^{r} [E_2 + r - D - A^{(i)} - k_i + \eta_i(2A^{(i)} + 2k_i + D + \beta + 1)]!}$$

$$(4.2.23)$$

PROOF. In order to confirm (4.2.21), in Theorem 17 set $E_1^{(i)} = E + 2 - i$ and $E_2^{(i)} = E + i$. Thus we obtain for the desired generating function the expression

$$\det \left(q^{s(A^{(s)} - A^{(t)})} \left(\sum_{j \geq 0} q^{j(j+\beta+\gamma+A^{(t)}-s+1)} \begin{bmatrix} -\beta \\ j \end{bmatrix} \begin{bmatrix} \beta + 2E + 2 - D \\ E + s - A^{(t)} - D - j \end{bmatrix} \right. \right.$$
$$\left. \left. - q^{(s-1)(2A^{(t)}+D+\beta+1)} \sum_{j \geq 0} q^{j(j+\beta+\gamma+A^{(t)}+s-1)} \begin{bmatrix} -\beta \\ j \end{bmatrix} \begin{bmatrix} \beta + 2E + 2 - D \\ E - s - A^{(t)} - D - j + 1 \end{bmatrix} \right) \right).$$

Again the summation indices in the t-th column are denoted by k_t. Next we use the linearity in the t-th column to turn the last expression into

$$\sum_{k_1,\ldots,k_r \geq 0} \prod_{i=1}^{r} q^{k_i(k_i+\beta+\gamma+A^{(i)}-i+1)} \begin{bmatrix} -\beta \\ k_i \end{bmatrix}$$
$$\det \left(q^{s(A^{(s)}+k_s-A^{(t)}-k_t)} \left(\begin{bmatrix} \beta + 2E + 2 - D \\ E + s - A^{(t)} - k_t - D \end{bmatrix} \right. \right.$$
$$\left. \left. - q^{(s-1)(2A^{(t)}+2k_t+D+\beta+1)} \begin{bmatrix} \beta + 2E + 2 - D \\ E - s - A^{(t)} - k_t - D + 1 \end{bmatrix} \right) \right).$$

The determinant in this last expression has already been evaluated. This is seen by replacing E by $E + \beta$, D by $D + \beta$, $A^{(t)}$ by $A^{(t)} + k_t$ in the proof of Theorem 8. The result is the expression (3.2.19), with these replacements. This furnishes (4.2.21) after having performed a little cancellation.

In order to establish (4.2.22), in (4.2.21) set $\beta = 1$, $D = 2D' + 1$, $\gamma = D' + 1$. Then it is seen that this just yields the expression (4.2.20) times some power of q. Hence the result is (4.2.18) times this power of q.

The proof of (4.2.23) proceeds in analogy with the proof of (3.2.20) and is therefore omitted. □

4.3. Tableaux generating functions. As a warm-up we provide the analogue of Theorem 9 for the case that the parts are not restricted to be odd. Unfortunately it is not as nice because it is not possible to evaluate the multiple sum which is obtained from (4.1.5).

Theorem 20. *The generating function* $\sum q^{n(\tau_1)+n(\tau_2)}$ *for pairs* (τ_1, τ_2) *of tableaux of identical shape consisting of at most* r *columns and with parts between 1 and* n *is given by*

$$\sum_{k_1,\ldots,k_r \geq 0} \prod_{i=1}^{r} (-1)^{k_i} q^{\binom{k_i+1}{2}} \frac{[2n+i]!}{[n+i+k_i]! \, [n+r-i-k_i]!} \prod_{1 \leq i < j \leq r} [j+k_j-i-k_i] \, . \quad (4.3.1)$$

In terms of Schur functions: The sum $\sum_{\lambda, \lambda_1 \leq r} s_\lambda^2(q^n, q^{n-1}, \ldots, q)$ *equals the expression* (4.3.1).

PROOF. By Proposition 28 we know that the pairs of tableaux in question are in one-to-one correspondence with families $\mathfrak{P} = (P_1, \ldots, P_r)$ of nonintersecting lattice

Consequently, by the definitions of ϑ_3 (cf. (4.2.3)) and of $\mathrm{ymaj}_{\beta;\gamma}^{(\eta)}$ (cf. (4.2.4)) we obtain

$$\vartheta_3(\Theta_3(\mathfrak{P}))$$
$$= (-1)^{\|\eta^{(1)}\|}\,\mathrm{sgn}\,\sigma\, q^{\sum_{i=1}^{r}\left(i(A^{(i)}-A^{(\sigma(i))})+i\eta_i^{(1)}(2A^{(\sigma(i))}+D+\beta+1)\right)}q^{\mathrm{ymaj}_{\beta;\gamma}^{(\eta^{(1)})}}\Theta_3(\mathfrak{P})$$
$$= -(-1)^{\|\eta\|}\,\mathrm{sgn}\,\sigma\, q^{\sum_{i=1}^{r}i(A^{(i)}-A^{(\sigma(i))})}q^{\sum_{i=1}^{r}i\eta_i(2A^{(\sigma(i))}+D+\beta+1)+(2A^{(\sigma(1))}+D+\beta+1)}$$
$$\cdot q^{\mathrm{ymaj}_{\beta;\gamma}^{(\eta)}\,\mathfrak{P}-\left(A^{(\sigma(1))}+D\right)-A^{(\sigma(1))}-1-\beta}$$
$$= -\vartheta_3(\mathfrak{P}). \tag{4.2.9}$$

Likewise, if $\eta_1 = 1$ by (3.2.5) and (5.2.2) we have

$$\mathrm{ymaj}_{\beta;\gamma}((\Phi_\gamma^{(1)})^{(-1)}(P_1)) = \mathrm{xmaj}_{\beta;\gamma+1}(P_1) + A^{(\sigma(1))} + D + A^{(\sigma(1))} + 1 + \beta.$$

Consequently, by the definitions (4.2.3) and (4.2.4) we obtain

$$\vartheta_3(\Theta_3(\mathfrak{P}))$$
$$= (-1)^{\|\eta^{(1)}\|}\,\mathrm{sgn}\,\sigma\, q^{\sum_{i=1}^{r}\left(i(A^{(i)}-A^{(\sigma(i))})+i\eta_i^{(1)}(2A^{(\sigma(i))}+D+\beta+1)\right)}q^{\mathrm{ymaj}_{\beta;\gamma}^{(\eta^{(1)})}}\Theta_3(\mathfrak{P})$$
$$= -(-1)^{\|\eta\|}\,\mathrm{sgn}\,\sigma\, q^{\sum_{i=1}^{r}i(A^{(i)}-A^{(\sigma(i))})}q^{\sum_{i=1}^{r}i\eta_i(2A^{(\sigma(i))}+D+\beta+1)-(2A^{(\sigma(1))}+D+\beta+1)}$$
$$\cdot q^{\mathrm{ymaj}_{\beta;\gamma}^{(\eta)}\,\mathfrak{P}+A^{(\sigma(1))}+D+A^{(\sigma(1))}+1+\beta}$$
$$= -\vartheta_3(\mathfrak{P}). \tag{4.2.10}$$

Now, since by (4.2.8), (4.2.9), (4.2.10) we finally settled that Θ_3 is sign-reversing with respect to ϑ_3, we may infer

$$\sum_{\sigma\in\mathfrak{S}_r,\,\eta\in\{0,1\}^r} GF(Q(\mathfrak{R}^\eta\mathbf{A}_\sigma \to \mathbf{E})^-;\vartheta_3) = 0. \tag{4.2.11}$$

All that remains, is to analogize the computation (3.2.9). Indeed, by consecutive use of (4.2.5), (4.2.11), (2.2.16), (5.1.11) and (5.1.12),

$$GF(Q(\mathbf{A} \to \mathbf{E})^+; \vartheta_3) = GF(Q(\mathbf{A} \to \mathbf{E}); \vartheta_3) - GF(Q(\mathbf{A} \to \mathbf{E})^-; \vartheta_3)$$

$$= GF(Q(\mathbf{A} \to \mathbf{E}); \vartheta_3) + \sum_{\substack{\sigma \in \mathfrak{S}_r, \eta \in \{0,1\}^r \\ (\sigma,\eta) \neq (id,\mathbf{0})}} GF(Q(\mathfrak{R}^\eta \mathbf{A}_\sigma \to \mathbf{E})^-; \vartheta_3)$$

$$= \sum_{\sigma \in \mathfrak{S}_r, \eta \in \{0,1\}^r} GF(Q(\mathfrak{R}^\eta \mathbf{A}_\sigma \to \mathbf{E}); \vartheta_3)$$

$$= \sum_{\sigma \in \mathfrak{S}_r} \operatorname{sgn} \sigma \prod_{i=1}^r q^{i(A^{(i)} - A^{(\sigma(i))})} (GF(P(\mathcal{A}_{\sigma(i)} \to \mathcal{E}_i); \operatorname{ymaj}_{\beta; \gamma - i + 1})$$

$$- q^{i(2A^{(\sigma(i))} + D + \beta + 1)} GF(P(\mathfrak{R}\mathcal{A}_{\sigma(i)} \to \mathcal{E}_i); \operatorname{xmaj}_{\beta; \gamma + i}))$$

$$= \sum_{\sigma \in \mathfrak{S}_r} \operatorname{sgn} \sigma \prod_{i=1}^r q^{i(A^{(i)} - A^{(\sigma(i))})}$$

$$\times \left(\sum_{j \geq 0} q^{j(j + \beta + \gamma - i + 1 + A^{(\sigma(i))})} \begin{bmatrix} -\beta \\ j \end{bmatrix} \begin{bmatrix} \beta + E_1^{(i)} + E_2^{(i)} - D \\ E_1^{(i)} - A^{(\sigma(i))} - D - j \end{bmatrix} \right.$$

$$- q^{i(2A^{(\sigma(i))} + D + \beta + 1)} \sum_{j \geq 0} q^{j(j + \beta + \gamma + i + A^{(\sigma(i))} + 1)}$$

$$\left. \begin{bmatrix} -\beta \\ j \end{bmatrix} \begin{bmatrix} \beta + E_1^{(i)} + E_2^{(i)} - D \\ E_2^{(i)} - A^{(\sigma(i))} - D - j - 1 \end{bmatrix} \right)$$

$$= \det_{1 \leq s, t \leq r} \left(q^{s(A^{(s)} - A^{(t)})} \left(\sum_{j \geq 0} q^{j(j + \beta + \gamma + A^{(t)} - s + 1)} \begin{bmatrix} -\beta \\ j \end{bmatrix} \begin{bmatrix} \beta + E_1^{(s)} + E_2^{(s)} - D \\ E_1^{(s)} - A^{(t)} - D - j \end{bmatrix} \right.$$

$$- q^{s(2A^{(t)} + D + \beta + 1)} \sum_{j \geq 0} q^{j(j + \beta + \gamma + A^{(t)} + s + 1)}$$

$$\left. \left. \begin{bmatrix} -\beta \\ j \end{bmatrix} \begin{bmatrix} \beta + E_1^{(s)} + E_2^{(s)} - D \\ E_2^{(s)} - A^{(t)} - D - j - 1 \end{bmatrix} \right) \right).$$

Note that because of the restrictions (4.2.1a), the requirements being necessary for being allowed to use (5.1.11) and (5.1.12) are always satisfied. The last line is identical with (4.2.2).   □

If the paths lie above the line $x = y$, there is a result for the xmaj.

Theorem 17. *Let* $\mathcal{A}_i = (-A^{(i)}, A^{(i)} + D)$ *and* $\mathcal{E}_i = (E_1^{(i)}, E_2^{(i)})$, $i = 1, 2, \ldots, r$, *be lattice points in the integer lattice* $\mathbb{Z}^2$ *such that (3.2.10) holds. Let* γ *be an integer* *satisfying*

$$\max_{1\le i\le r}(D - E_2^{(i)} + i - 1) \le \gamma \le \min_{1\le i\le r}(E_1^{(i)} + i - 1),$$

$$\max_{1\le i\le r}(D - E_1^{(i)} - i + 2) \le \gamma \le \min_{1\le i\le r}(E_2^{(i)} - i + 2),$$

$$\text{and} \quad -A^{(1)} \le \gamma \le A^{(1)} + D. \quad (4.2.12)$$

The generating function $\sum q^{\mathrm{xmaj}_{\beta;\gamma}(\mathfrak{P})}$, *where the sum is over all nonintersecting families* $\mathfrak{P} = (P_1, \ldots, P_r)$ *of lattice paths which lie above the line* $x = y$, $P_i : \mathcal{A}_i \to \mathcal{E}_i$, $i = 1, 2, \ldots, r$, *and where* $\mathrm{xmaj}_{\beta;\gamma}\mathfrak{P}$ *is given by (4.0.9) (see the Remark below), is equal to the expression*

$$\det_{1\le s,t\le r}\left(q^{s(A^{(s)} - A^{(t)})}\left(\sum_{j\ge 0} q^{j(j+\beta+\gamma+A^{(t)}-s+1)} \begin{bmatrix} -\beta \\ j \end{bmatrix} \begin{bmatrix} \beta + E_1^{(s)} + E_2^{(s)} - D \\ E_2^{(s)} - A^{(t)} - D - j \end{bmatrix} \right.$$

$$\left. - q^{(s-1)(2A^{(t)}+D+\beta+1)} \sum_{j\ge 0} q^{j(j+\beta+\gamma+A^{(t)}+s-1)} \begin{bmatrix} -\beta \\ j \end{bmatrix} \begin{bmatrix} \beta + E_1^{(s)} + E_2^{(s)} - D \\ E_1^{(s)} - A^{(t)} - D - j - 1 \end{bmatrix} \right) \right).$$

$$(4.2.13)$$

REMARK. Because of (3.2.10a,b) the paths $P_1, \ldots, P_r$ are ordered "from bottom to top". It is important to observe that since we made the definition (4.0.9) for $\mathrm{xmaj}_{\beta;\gamma}$ of a family of paths, by requiring (2.2.2), dependent on the order of the paths, here in fact we have

$$\mathrm{xmaj}_{\beta;\gamma}(\mathfrak{P}) = \sum_{i=1}^{r} \mathrm{xmaj}_{\beta;\gamma-i+1}(P_i). \quad (4.2.14)$$

SKETCH OF PROOF. We follow the line of the proof of Theorem 16. The main difference is that instead of $\Phi_\gamma^{(1)}$ we have to use $\Phi^{(2)}$, which is defined in Proposition 26. Of course, everywhere the roles of ymaj and xmaj have to be exchanged. We only remark that the weight function to be used is

$$\vartheta_4(\mathfrak{P}) = (-1)^{\|\eta\|} \operatorname{sgn}\sigma \, q^{\sum_{i=1}^r \left(i(A^{(i)} - A^{(\sigma(i))}) + i\eta_i(2A^{(\sigma(i))} + D + \beta + 1) \right)} q^{\mathrm{xmaj}_{\beta;\gamma}^{(\eta)}\mathfrak{P}}$$

$$\text{if } \mathfrak{P} \in Q(\mathfrak{R}^\eta \mathbf{A}_\sigma \to \mathbf{E}), \quad (4.2.15)$$

where $\mathrm{xmaj}_{\beta;\gamma}^{(\eta)}$ is given by

$$\mathrm{xmaj}_{\beta;\gamma}^{(\eta)}(\mathfrak{P}) = \sum_{i=1}^{r} \left((1 - \eta_i)\,\mathrm{xmaj}_{\beta;\gamma-i+1}(P_i) + \eta_i(\mathrm{ymaj}_{\beta;\gamma+i-2}(P_i)) \right). \quad (4.2.16)$$

What has to be determined, is the generating function $GF(\bar{Q}(\mathbf{A} \to \mathbf{E})^+, \vartheta_4)$, where $\bar{Q}(\mathbf{A} \to \mathbf{E})^+$ is the set of all families $\mathfrak{P} = (P_1, \ldots, P_r)$ of nonintersecting lattice paths which lie *above* $x = y$, $P_i : \mathcal{A}_i \to \mathcal{E}_i$. This is done quite analogously as in the proof of Theorem 16 and is therefore left to the reader.   □

Next we consider cases in which (4.2.2) or (4.2.13) can be simplified. With the special final points of Theorem 7, we can get rid of the determinant in (4.2.2) thus obtaining the multiple sum (4.2.17) and for the more general case the multiple sum (4.2.19). If $\beta = 1$ and $\gamma = (D-1)/2$, with the help of Proposition 36 the sum (4.2.17) can even be evaluated to result in a nice product. Since γ and D are integers, the relation $\gamma = (D-1)/2$ implies that in this case D has to be odd. There do not seem to be other special cases where (4.2.17) or (4.2.19) can be evaluated. In particular, for $r = 1$ none of the known basic hypergeometric summation theorems can be applied to (4.2.17) except for $\beta = 1$ and $\gamma = (D-1)/2$. So it is very unlikely that (4.2.17) can be summed for $r > 1$ in other cases.

Theorem 18. *Let* $\mathcal{A}_i = (A^{(i)} + D, -A^{(i)})$ *and* $\mathcal{E}_i = (E+i, E+2-i)$, $i = 1, 2, \ldots, r$, *be lattice points in the integer lattice* $\mathbb{Z}^2$ *such that (3.1.1a) and*

$$2A^{(i)} + D \geq 0, \quad i = 1, 2, \ldots, r,$$

hold. If γ *is an integer satisfying*

$$D - E - 1 \leq \gamma \leq E \quad \text{and} \quad -A^{(1)} \leq \gamma \leq A^{(1)} + D,$$

then the generating function $\sum q^{\mathrm{ymaj}_{\beta;\gamma}(\mathfrak{P})}$ *where the sum is over all nonintersecting families* $\mathfrak{P} = (P_1, \ldots, P_r)$ *of lattice paths which lie below the line* $x = y$, $P_i : \mathcal{A}_i \to \mathcal{E}_i$, $i = 1, 2, \ldots, r$, *is equal to the expression*

$$\sum_{k_1, \ldots, k_r \geq 0} \prod_{i=1}^r q^{k_i(k_i + \beta + \gamma + A^{(i)} - i + 1)} \begin{bmatrix} -\beta \\ k_i \end{bmatrix}$$

$$\frac{[\beta + 2E + 2i - D]!}{[E + r - A^{(i)} - k_i - D]! \, [\beta + E + r + 1 + A^{(i)} + k_i]!}$$

$$\prod_{1 \leq i < j \leq r} [A^{(j)} + k_j - A^{(i)} - k_i] \prod_{1 \leq i \leq j \leq r} [A^{(i)} + k_i + A^{(j)} + k_j + D + \beta + 1] . \quad (4.2.17)$$

In particular, if D is odd, say $D = 2D' + 1$, and if $-A^{(1)} \leq D' \leq E$, the generating function $\sum q^{\mathrm{ymaj}_{1;D'}(\mathfrak{P})}$ over the same set of nonintersecting families $\mathfrak{P}$ is given by

$$\prod_{i=1}^r \frac{[2E + 2i - 2D']!}{[E + r - A^{(i)} - 2D' - 1]! \, [E + r + 1 + A^{(i)}]!} \prod_{1 \leq i < j \leq r} [A^{(j)} - A^{(i)}]$$

$$\times \prod_{1 \leq i < j \leq r} [A^{(i)} + A^{(j)} + 2D' + 2] \prod_{i=1}^r \frac{[A^{(i)} + D' + 1]}{[E + i - D']} . \quad (4.2.18)$$

More generally, if D is again generic and if the final points of the lattice paths are $\mathcal{E}_i = (E_1 + i, E_2 - i)$, $i = 1, 2, \ldots, r$, with $E_1 \geq E_2 - 2$, then the generating function

$\sum q^{\mathrm{ymaj}_{\beta,\gamma}}(\mathfrak{P})$ for all nonintersecting families $\mathfrak{P} = (P_1, \ldots, P_r)$ of lattice paths which lie below the line $x = y$, $P_i : \mathcal{A}_i \to \mathcal{E}_i$, $i = 1, 2, \ldots, r$, is equal to the expression

$$\sum_{\eta \in \{0,1\}^r} (-1)^{\|\eta\|} \sum_{k_1, \ldots, k_r \geq 0} \prod_{i=1}^r \left(q^{i\eta_i(2A^{(i)}+2k_i+D+\beta+1)+k_i(k_i+\beta+\gamma+A^{(i)}-i+1)} \begin{bmatrix} -\beta \\ k_i \end{bmatrix} \right.$$

$$\left. \frac{[\beta + E_1 + E_2 + i - D - 1]!}{[\beta + E_2 + A^{(i)} + k_i - \eta_i(2A^{(i)} + 2k_i + D + \beta + 1) - 1]!} \right)$$

$$\frac{\prod_{1 \leq i < j \leq r} [A^{(j)} + k_j - \eta_j(2A^{(j)} + 2k_j + D + \beta + 1) - A^{(i)} - k_i + \eta_i(2A^{(i)} + 2k_i + D + \beta + 1)]}{\prod_{i=1}^r [E_1 + r - D - A^{(i)} - k_i + \eta_i(2A^{(i)} + 2k_i + D + \beta + 1)]!}$$

$$(4.2.19)$$

PROOF. In order to confirm (4.2.17), in Theorem 16 set $E_1^{(i)} = E + i$ and $E_2^{(i)} = E + 2 - i$ to obtain for the desired generating function the expression

$$\det \left(q^{s(A^{(s)} - A^{(t)})} \left(\sum_{j \geq 0} q^{j(j+\beta+\gamma+A^{(t)}-s+1)} \begin{bmatrix} -\beta \\ j \end{bmatrix} \begin{bmatrix} \beta + 2E + 2 - D \\ E + s - A^{(t)} - D - j \end{bmatrix} \right. \right.$$

$$\left. \left. - q^{s(2A^{(t)}+D+\beta+1)} \sum_{j \geq 0} q^{j(j+\beta+\gamma+A^{(t)}+s+1)} \begin{bmatrix} -\beta \\ j \end{bmatrix} \begin{bmatrix} \beta + 2E + 2 - D \\ E - s - A^{(t)} - D - j + 1 \end{bmatrix} \right) \right).$$

As before in the proof of Theorem 15, we denote the summation indices in the t-th column by k_t and then use the linearity in the t-th column to obtain that the last expression equals

$$\sum_{k_1, \ldots, k_r \geq 0} \prod_{i=1}^r q^{k_i(k_i+\beta+\gamma+A^{(i)}-i+1)} \begin{bmatrix} -\beta \\ k_i \end{bmatrix}$$

$$\det \left(q^{s(A^{(s)}+k_s-A^{(t)}-k_t)} \left(\begin{bmatrix} \beta + 2E + 2 - D \\ E + s - A^{(t)} - k_t - D \end{bmatrix} \right. \right.$$

$$\left. \left. - q^{s(2A^{(t)}+2k_t+D+\beta+1)} \begin{bmatrix} \beta + 2E + 2 - D \\ E - s - A^{(t)} - k_t - D + 1 \end{bmatrix} \right) \right).$$

The determinant in this last expression has already been evaluated. This is seen by replacing E by $E + \beta$, D by $D + \beta$, $A^{(t)}$ by $A^{(t)} + k_t$ in the proof of Theorem 7. The result is the expression (3.2.14), with these replacements. This immediately settles (4.2.17).

In order to establish (4.2.18), in (4.2.17) set $\beta = 1$, $D = 2D' + 1$, $\gamma = D'$. Use hypergeometric notation to obtain for the desired generating function the expression

$$\prod_{i=1}^{r} \frac{[2E+2i-2D']!}{[E+r-A^{(i)}-2D'-1]![E+r+2+A^{(i)}]!}$$

$$\times \sum_{k_1,\ldots,k_r \geq 0} \prod_{i=1}^{r} q^{k_i(1+E-D'+r-i)} \frac{(q^{A^{(i)}+2D'+1-E-r})_{k_i}}{(q^{E+r+3+A^{(i)}})_{k_i}}$$

$$\prod_{1\leq i<j\leq r} (1-q^{A^{(j)}+k_j-A^{(i)}-k_i}) \prod_{1\leq i\leq j\leq r} (1-q^{A^{(i)}+k_i+A^{(j)}+k_j+2D'+3}) . \quad (4.2.20)$$

By setting $m_i = q^{A^{(i)}+D'+3/2}$, $i = 1,2,\ldots,r$, and $A = q^{D'-E-r-1/2}$, this can be written in the form

$$\prod_{i=1}^{r} \frac{[2E+2i-2D']!}{[E+r-A^{(i)}-2D'-1]![E+r+2+A^{(i)}]!}$$

$$\times \prod_{1\leq i<j\leq r} [A^{(j)}-A^{(i)}] \prod_{1\leq i\leq j\leq r} [A^{(i)}+A^{(j)}+2D'+3]$$

$$\times \sum_{k_1,\ldots,k_r \geq 0} \prod_{i=1}^{r} \left(\frac{\sqrt{q}}{q^i A}\right)^{k_i} \prod_{i=1}^{r} \frac{(m_i A)_{k_i}}{(qm_i/A)_{k_i}}$$

$$\times \prod_{1\leq i<j\leq r} \frac{1-\frac{m_j}{m_i}q^{k_j-k_i}}{1-\frac{m_j}{m_i}} \prod_{1\leq i\leq j\leq r} \frac{1-m_i m_j q^{k_i+k_j}}{1-m_i m_j} .$$

The r-fold sum is evaluated using (5.6.1). Thus we obtain

$$\prod_{i=1}^{r} \frac{[2E+2i-2D']!}{[E+r-A^{(i)}-2D'-1]![E+r+2+A^{(i)}]!}$$

$$\times \prod_{1\leq i<j\leq r} [A^{(j)}-A^{(i)}] \prod_{1\leq i\leq j\leq r} [A^{(i)}+A^{(j)}+2D'+3]$$

$$\times \prod_{1\leq i<j\leq r} \frac{[A^{(i)}+A^{(j)}+2D'+2]}{[A^{(i)}+A^{(j)}+2D'+3]} \prod_{i=1}^{r} \left(\frac{[A^{(i)}+D'+1]}{[2A^{(i)}+2D'+3]} \frac{[E+A^{(i)}+r+2]}{[E+r-D'-i+1]}\right),$$

which simplifies to (4.2.18).

The proof of (4.2.19) proceeds in analogy with the proof of (3.2.15) and is therefore omitted. □

The analogue of Theorem 18 which corresponds to Theorem 17 is the following result. A closed form is obtained for $E_1^{(i)} = E+2-i$, $E_2^{(i)} = E+i$, $\beta = 1$ and $\gamma = (D+1)/2$, which forces D to be odd here, too. Again, obtaining a closed form for other choices of the parameters seems to be very unlikely.

Theorem 19. *Let $\mathcal{A}_i = (-A^{(i)}, A^{(i)} + D)$ and $\mathcal{E}_i = (E+2-i, E+i)$, $i = 1, 2, \ldots, r$, be lattice points in the integer lattice $\mathbb{Z}^2$ such that (3.2.10a) and*

$$2A^{(i)} + D \geq 0, \quad i = 1, 2, \ldots, r,$$

hold. If γ is an integer satisfying

$$D - E \leq \gamma \leq E + 1 \quad \text{and} \quad -A^{(1)} \leq \gamma \leq A^{(1)} + D,$$

then the generating function $\sum q^{\mathrm{xmaj}_{\beta;\gamma}}(\mathfrak{P})$ where the sum is over all nonintersecting families $\mathfrak{P} = (P_1, \ldots, P_r)$ of lattice paths which lie above the line $x = y$, $P_i : \mathcal{A}_i \to \mathcal{E}_i$, $i = 1, 2, \ldots, r$ is equal to the expression

$$q^{\binom{r+1}{2} + r(E-D) - \sum_{i=1}^{r} A^{(i)}} \sum_{k_1, \ldots, k_r \geq 0} \prod_{i=1}^{r} q^{k_i(k_i + \beta + \gamma + A^{(i)} - i)}$$

$$\begin{bmatrix} -\beta \\ k_i \end{bmatrix} \frac{[\beta + 2E + 2i - D]!}{[E + r - A^{(i)} - k_i - D]! \, [\beta + E + r + 1 + A^{(i)} + k_i]!}$$

$$\prod_{1 \leq i < j \leq r} [A^{(j)} + k_j - A^{(i)} - k_i] \prod_{1 \leq i \leq j \leq r} [A^{(i)} + k_i + A^{(j)} + k_j + D + \beta + 1] . \quad (4.2.21)$$

In particular, if D is odd, say $D = 2D' + 1$, and if $-A^{(1)} \leq D' \leq E$, the generating function $\sum q^{\mathrm{xmaj}_{1;D'+1}}(\mathfrak{P})$ over the same set of nonintersecting families $\mathfrak{P}$ is given by

$$q^{\binom{r+1}{2} + r(E-D) - \sum_{i=1}^{r} A^{(i)}} \prod_{i=1}^{r} \frac{[2E + 2i - 2D']!}{[E + r - A^{(i)} - 2D' - 1]! \, [E + r + 1 + A^{(i)}]!}$$

$$\times \prod_{1 \leq i < j \leq r} [A^{(j)} - A^{(i)}] \prod_{1 \leq i \leq j \leq r} [A^{(i)} + A^{(j)} + 2D' + 2] \prod_{i=1}^{r} \frac{[A^{(i)} + D' + 1]}{[E + i - D']} . \quad (4.2.22)$$

More generally, if D is again generic and if the final points of the lattice paths are $\mathcal{E}_i = (E_1 - i, E_2 + i)$, $i = 1, 2, \ldots, r$, with $E_1 \leq E_2 + 2$, then the generating function $\sum q^{\mathrm{xmaj}_{\beta;\gamma}}(\mathfrak{P})$ for all nonintersecting families $\mathfrak{P} = (P_1, \ldots, P_r)$ of lattice paths which lie above the line $x = y$, $P_i : \mathcal{A}_i \to \mathcal{E}_i$, $i = 1, 2, \ldots, r$, is equal to the expression

$$\sum_{\eta \in \{0,1\}^r} (-1)^{\|\eta\|} \sum_{k_1, \ldots, k_r \geq 0} \prod_{i=1}^{r} \left(q^{(i-1)\eta_i(2A^{(i)} + 2k_i + D + \beta + 1) + k_i(k_i + \beta + \gamma + A^{(i)} - i + 1)} \begin{bmatrix} -\beta \\ k_i \end{bmatrix} \right.$$

$$\left. \frac{[\beta + E_1 + E_2 + i - D - 1]!}{[\beta + E_1 + A^{(i)} + k_i - \eta_i(2A^{(i)} + 2k_i + D + \beta + 1) - 1]!} \right)$$

$$\frac{\prod_{1 \leq i < j \leq r} [A^{(j)} + k_j - \eta_j(2A^{(j)} + 2k_j + D + \beta + 1) - A^{(i)} - k_i + \eta_i(2A^{(i)} + 2k_i + D + \beta + 1)]}{\prod_{i=1}^{r} [E_2 + r - D - A^{(i)} - k_i + \eta_i(2A^{(i)} + 2k_i + D + \beta + 1)]!}$$

$$(4.2.23)$$

PROOF. In order to confirm (4.2.21), in Theorem 17 set $E_1^{(i)} = E + 2 - i$ and $E_2^{(i)} = E + i$. Thus we obtain for the desired generating function the expression

$$
\det \left(q^{s(A^{(s)} - A^{(t)})} \Big(\sum_{j \geq 0} q^{j(j+\beta+\gamma+A^{(t)}-s+1)} \begin{bmatrix} -\beta \\ j \end{bmatrix} \begin{bmatrix} \beta + 2E + 2 - D \\ E + s - A^{(t)} - D - j \end{bmatrix} \right.
$$
$$
\left. -q^{(s-1)(2A^{(t)}+D+\beta+1)} \sum_{j \geq 0} q^{j(j+\beta+\gamma+A^{(t)}+s-1)} \begin{bmatrix} -\beta \\ j \end{bmatrix} \begin{bmatrix} \beta + 2E + 2 - D \\ E - s - A^{(t)} - D - j + 1 \end{bmatrix} \Big) \right).
$$

Again the summation indices in the t-th column are denoted by k_t. Next we use the linearity in the t-th column to turn the last expression into

$$
\sum_{k_1, \ldots, k_r \geq 0} \prod_{i=1}^{r} q^{k_i(k_i+\beta+\gamma+A^{(i)}-i+1)} \begin{bmatrix} -\beta \\ k_i \end{bmatrix}
$$
$$
\det \left(q^{s(A^{(s)}+k_s-A^{(t)}-k_t)} \Big(\begin{bmatrix} \beta + 2E + 2 - D \\ E + s - A^{(t)} - k_t - D \end{bmatrix} \right.
$$
$$
\left. - q^{(s-1)(2A^{(t)}+2k_t+D+\beta+1)} \begin{bmatrix} \beta + 2E + 2 - D \\ E - s - A^{(t)} - k_t - D + 1 \end{bmatrix} \Big) \right).
$$

The determinant in this last expression has already been evaluated. This is seen by replacing E by $E + \beta$, D by $D + \beta$, $A^{(t)}$ by $A^{(t)} + k_t$ in the proof of Theorem 8. The result is the expression (3.2.19), with these replacements. This furnishes (4.2.21) after having performed a little cancellation.

In order to establish (4.2.22), in (4.2.21) set $\beta = 1$, $D = 2D' + 1$, $\gamma = D' + 1$. Then it is seen that this just yields the expression (4.2.20) times some power of q. Hence the result is (4.2.18) times this power of q.

The proof of (4.2.23) proceeds in analogy with the proof of (3.2.20) and is therefore omitted.   □

4.3. Tableaux generating functions. As a warm-up we provide the analogue of Theorem 9 for the case that the parts are not restricted to be odd. Unfortunately it is not as nice because it is not possible to evaluate the multiple sum which is obtained from (4.1.5).

Theorem 20. *The generating function* $\sum q^{n(\tau_1)+n(\tau_2)}$ *for pairs* (τ_1, τ_2) *of tableaux of identical shape consisting of at most r columns and with parts between 1 and n is given by*

$$
\sum_{k_1, \ldots, k_r \geq 0} \prod_{i=1}^{r} (-1)^{k_i} q^{\binom{k_i+1}{2}} \frac{[2n+i]!}{[n+i+k_i]! \, [n+r-i-k_i]!} \prod_{1 \leq i < j \leq r} [j + k_j - i - k_i] . \quad (4.3.1)
$$

In terms of Schur functions: The sum $\sum_{\lambda, \lambda_1 \leq r} s_\lambda^2(q^n, q^{n-1}, \ldots, q)$ *equals the expression* (4.3.1).

PROOF. By Proposition 28 we know that the pairs of tableaux in question are in one-to-one correspondence with families $\mathfrak{P} = (P_1, \ldots, P_r)$ of nonintersecting lattice

i by $k - j$ and interchange the order of summations. This transforms the last sum into

$$\sum_{j \geq 0} q^{j(j+\beta+\gamma-A_2)} \begin{bmatrix} E_2 - \gamma \\ j \end{bmatrix} \sum_{k \geq 0} q^{k(k-j)} \begin{bmatrix} E_1 - A_1 \\ k \end{bmatrix} \begin{bmatrix} \gamma - A_2 \\ k - j \end{bmatrix} .$$

The inner sum can be evaluated by the q-Vandermonde summation (see e.g. [1, (3.3.10)]), which we restate in the form

$$\sum_{K \geq 0} q^{K(K-H+M)} \begin{bmatrix} N \\ K \end{bmatrix} \begin{bmatrix} M \\ H - K \end{bmatrix} = \begin{bmatrix} N + M \\ H \end{bmatrix} . \tag{5.1.4}$$

Utilizing $\begin{bmatrix} \gamma - A_2 \\ k - j \end{bmatrix} = \begin{bmatrix} \gamma - A_2 \\ \gamma - A_2 + j - k \end{bmatrix}$ and (5.1.4) with $N = E_1 - A_1$, $M = \gamma - A_2$, $H = \gamma - A_2 + j$, we get

$$\sum_{P: \mathcal{A} \to \mathcal{E}} q^{\mathbf{ymaj}_{\beta;\gamma}(P)}$$

$$= \sum_{j \geq 0} q^{j(j+\beta+\gamma-A_2)} \begin{bmatrix} E_2 - \gamma \\ j \end{bmatrix} \begin{bmatrix} E_1 + \gamma - A_1 - A_2 \\ j + \gamma - A_2 \end{bmatrix}$$

$$= \sum_{j \geq 0} q^{j(j+\beta+\gamma-A_2)} \begin{bmatrix} E_2 - \gamma \\ j \end{bmatrix} \begin{bmatrix} E_1 + \gamma - A_1 - A_2 \\ E_1 - A_1 - j \end{bmatrix} , \tag{5.1.5}$$

provided $A_2 \leq \gamma \leq E_2$.

There is an alternative expression for the ymaj-generating function which is obtained using Heine's basic hypergeometric transformation formula (cf. [14, Appendix (III.2)])

$${}_2\phi_1 \begin{bmatrix} a, b \\ c \end{bmatrix} ; q, z \end{bmatrix} = \frac{(c/b; q)_\infty (bz; q)_\infty}{(c; q)_\infty (z; q)_\infty} {}_2\phi_1 \begin{bmatrix} abz/c, b \\ bz \end{bmatrix} ; q, \frac{c}{b} \end{bmatrix} . \tag{5.1.6}$$

First we have to write the sum in the last line of (5.1.5) in hypergeometric notation (cf. subsection 2.1),

$$\sum_{P: \mathcal{A} \to \mathcal{E}} q^{\mathbf{ymaj}_{\beta;\gamma}(P)} = \frac{(q^{E_1-A_1+1})_{\gamma-A_2}}{(q)_{\gamma-A_2}} {}_2\phi_1 \begin{bmatrix} q^{\gamma-E_2}, q^{A_1-E_1} \\ q^{\gamma-A_2+1} \end{bmatrix} ; q, q^{1+\beta+E_1+E_2-A_1-A_2} \end{bmatrix} .$$

Now (5.1.6) can be applied to get

$$\sum_{P: \mathcal{A} \to \mathcal{E}} q^{\mathbf{ymaj}_{\beta;\gamma}(P)} = \frac{(q^{E_1-A_1+1})_{\gamma-A_2} (q^{\gamma+E_1-A_1-A_2+1})_\infty (q^{1+\beta+E_2-A_2})_\infty}{(q)_{\gamma-A_2} (q^{\gamma-A_2+1})_\infty (q^{1+\beta+E_1+E_2-A_1-A_2})_\infty}$$

$$ \qquad {}_2\phi_1 \begin{bmatrix} q^\beta, q^{A_1-E_1} \\ q^{1+\beta+E_2-A_2} \end{bmatrix} ; q, q^{1+\gamma+E_1-A_1-A_2} \end{bmatrix} ,$$

and finally, after simplifying and rewriting this in q-binomial notation,

$$\sum_{P: \mathcal{A} \to \mathcal{E}} q^{\mathbf{ymaj}_{\beta;\gamma}(P)} = \sum_{j > 0} q^{j(j+\beta+\gamma-A_2)} \begin{bmatrix} -\beta \\ j \end{bmatrix} \begin{bmatrix} \beta + E_1 + E_2 - A_1 - A_2 \\ E_1 - A_1 - j \end{bmatrix} , \tag{5.1.7}$$

provided $A_2 \leq \gamma \leq E_2$.

Now we turn to the xmaj-generating function. Let $A_1 \leq \gamma \leq E_1$. By (4.0.7) we have

$$\sum_{P:\mathcal{A}\to\mathcal{E}} q^{\mathrm{xmaj}_{\beta;\gamma}(P)}$$

$$= \sum_{k\geq 0} \sum_{\substack{a\in P_k[A_1,E_1-1] \\ b\in P_k[A_2+1,E_2]}} q^{\|a\|+\|b\|+\beta\cdot|\{t:a_t\geq\gamma\}|-k(A_1+A_2)}$$

$$= \sum_{k\geq 0} \sum_{\substack{a'\in P_k[A_1+1,E_1] \\ b'\in P_k[A_2,E_2-1]}} q^{\|a'\|+\|b'\|+\beta\cdot|\{t:a'_t>\gamma\}|-k(A_1+A_2)} ,$$

where $a'_i := a_i + 1$ and $b'_i = b_i - 1$, $i = 1, \ldots, k$. Comparing the last line with (5.1.3), one observes that to obtain expressions for the xmaj-generating function in the previous computations for the ymaj only the roles of A_1 and A_2, and those of E_1 and E_2 have to be exchanged. Therefore, from (5.1.5) we get

$$\sum_{P:\mathcal{A}\to\mathcal{E}} q^{\mathrm{xmaj}_{\beta;\gamma}(P)} = \sum_{j\geq 0} q^{j(j+\beta+\gamma-A_1)} \begin{bmatrix} E_1 - \gamma \\ j \end{bmatrix} \begin{bmatrix} E_2 + \gamma - A_1 - A_2 \\ E_2 - A_2 - j \end{bmatrix} , \qquad (5.1.8)$$

and from (5.1.7) we obtain

$$\sum_{P:\mathcal{A}\to\mathcal{E}} q^{\mathrm{xmaj}_{\beta;\gamma}(P)} = \sum_{j\geq 0} q^{j(j+\beta+\gamma-A_1)} \begin{bmatrix} -\beta \\ j \end{bmatrix} \begin{bmatrix} \beta + E_1 + E_2 - A_1 - A_2 \\ E_2 - A_2 - j \end{bmatrix} , \qquad (5.1.9)$$

both identities holding provided $A_1 \leq \gamma \leq E_1$.

There is an important connection between the ymaj- and the xmaj-generating function. This is revealed by combining (5.1.5) and (5.1.8). Let $A_1+A_2-E_1 \leq \gamma < A_2$ and in (5.1.8) replace γ by $A_1 + A_2 - \gamma$. Then shift the summation index by replacing j by $j+\gamma-A_2$:

$$\sum_{P:\mathcal{A}\to\mathcal{E}} q^{\mathrm{xmaj}_{\beta;A_1+A_2-\gamma}(P)} = \sum_{j\geq A_2-\gamma} q^{(j+\gamma-A_2)(j+\beta)} \begin{bmatrix} E_1 + \gamma - A_1 - A_2 \\ j + \gamma - A_2 \end{bmatrix} \begin{bmatrix} E_2 - \gamma \\ E_2 - \gamma - j \end{bmatrix}$$

$$= q^{\beta(\gamma-A_2)} \sum_{j\geq 0} q^{j(j+\beta+\gamma-A_2)} \begin{bmatrix} E_2 - \gamma \\ j \end{bmatrix} \begin{bmatrix} E_1 + \gamma - A_1 - A_2 \\ E_1 - A_1 - j \end{bmatrix} . \qquad (5.1.10)$$

Comparing this last line with (5.1.5) leads to the surprising discovery that the sum in (5.1.5), and hence also the sum in (5.1.7), for $\gamma \geq A_2$ is equal to the ymaj$_{\beta;\gamma}$-generating function for paths from $\mathcal{A}$ to $\mathcal{E}$, while for $\gamma < A_2$ both sums are, save for the factor $q^{\beta(\gamma-A_2)}$, equal to the xmaj$_{\beta;A_1+A_2-\gamma}$-generating function for the same family of paths. Similarly, replacing γ by $A_1 + A_2 - \gamma$ in the computation (5.1.10) and in (5.1.5) shows that the sum in (5.1.8), and hence also the sum in (5.1.9), for $\gamma \geq A_1$ is equal to the xmaj$_{\beta;\gamma}$-generating function for paths from $\mathcal{A}$ to $\mathcal{E}$, while for $\gamma < A_1$ both sums are, save to the factor $q^{\beta(\gamma-A_1)}$, equal to the ymaj$_{\beta;A_1+A_2-\gamma}$-generating function for the same family of paths. It is these observations that justify the seemingly artificial extension (4.0.3) of ymaj$_{\beta;\gamma}$ to $\gamma < A_2$ and the likewise seemingly artificial extension (4.0.4) of xmaj$_{\beta;\gamma}$ to $\gamma < A_1$.

The next Lemma summarizes results about ymaj- and xmaj-generating functions for unrestricted paths. Most of them have already been proved.

Lemma 23. *Let* $\mathcal{A} = (A_1, A_2)$ *and* $\mathcal{E} = (E_1, E_2)$ *be two lattice points in the integer lattice* $\mathbb{Z}^2$.

For $A_1 + A_2 - E_1 \leq \gamma \leq E_2$ *the* ymaj$_{\beta;\gamma}$-*generating function for the paths from* $\mathcal{A}$ *to* $\mathcal{E}$ *is given by*

$$\sum_{P:\mathcal{A}\to\mathcal{E}} q^{\text{ymaj}_{\beta;\gamma}(P)} = \sum_{j\geq 0} q^{j(j+\beta+\gamma-A_2)} \begin{bmatrix} -\beta \\ j \end{bmatrix} \begin{bmatrix} \beta + E_1 + E_2 - A_1 - A_2 \\ E_1 - A_1 - j \end{bmatrix}. \quad (5.1.11)$$

For $A_1 + A_2 - E_2 \leq \gamma \leq E_1$ *the* xmaj$_{\beta;\gamma}$-*generating function for the paths from* $\mathcal{A}$ *to* $\mathcal{E}$ *is given by*

$$\sum_{P:\mathcal{A}\to\mathcal{E}} q^{\text{xmaj}_{\beta;\gamma}(P)} = \sum_{j\geq 0} q^{j(j+\beta+\gamma-A_1)} \begin{bmatrix} -\beta \\ j \end{bmatrix} \begin{bmatrix} \beta + E_1 + E_2 - A_1 - A_2 \\ E_2 - A_2 - j \end{bmatrix}. \quad (5.1.12)$$

In particular, the maj-*generating function is given by*

$$\sum_{P:\mathcal{A}\to\mathcal{E}} q^{\text{maj}\, P} = \begin{bmatrix} E_1 + E_2 - A_1 - A_2 \\ E_1 - A_1. \end{bmatrix} \quad (5.1.13)$$

REMARK. Of course (5.1.13) is a well-known result of MacMahon ([32, p. 206], cf. [1, Theorem 3.1]).

PROOF OF THE LEMMA. First let $\mathcal{A} \leq \mathcal{E}$. That (5.1.11) holds for $A_2 \leq \gamma \leq E_2$ has been shown above when establishing (5.1.7) via (5.1.5). In case that $A_1 + A_2 - E_1 \leq \gamma < A_2$, by (4.0.3) ymaj$_{\beta;\gamma}$ is equal to xmaj$_{\beta;A_1+A_2-\gamma} + \beta(A_2 - \gamma)$. Because of $A_1 < A_1 + A_2 - \gamma \leq E_1$, use of (5.1.10) and the transformation that leads from (5.1.5) to (5.1.7) shows that (5.1.11) also holds in this case. The arguments for (5.1.12) are similar. (5.1.13) can be easily derived from (5.1.11) or (5.1.12) by setting $\beta = 0$ and applying the q-Vandermonde summation (5.1.4).

What remains is to show that (5.1.11)–(5.1.13) also hold for $\mathcal{A} \not\leq \mathcal{E}$. Since in this case there are no paths between $\mathcal{A}$ and $\mathcal{E}$, we have to show that the right-hand sides in (5.1.11)–(5.1.13) equal zero. This is trivial for (5.1.13). Next we turn to (5.1.11). If $E_1 < A_1$ all the summands in (5.1.11) are zero because of the second q-binomial coefficient. Now suppose that $E_1 \geq A_1$ and $E_2 < A_2$. By a careful examination of the derivation of (5.1.7) from (5.1.5) by means of Heine's transformation formula, it is seen that the right-hand side of (5.1.11) agrees with the right-hand side of (5.1.5) for all choices of the parameters provided that $\gamma \geq A_2$ and provided that $E_1 - A_1$ is a nonnegative integer. Since because of the second requirement the summations are finite, both, (5.1.5) and (5.1.11), can be written as polynomials in q^γ. There are infinitely many values of γ (and therefore of q^γ) where (5.1.5) and (5.1.11) agree, hence they agree as polynomials in q^γ. This implies that they agree for all γ. But the summands in (5.1.5) for $\gamma \leq E_2$ are non-zero only if $j \leq E_2 - \gamma$ and $j \geq A_2 - \gamma$.

Supposing that such a j exists we would get $A_2 - \gamma \leq E_2 - \gamma$ in contradiction to the assumption $E_2 < A_2$. Therefore the sum in (5.1.5) equals zero for $E_1 \geq A_1$ and $E_2 < A_2$, hence also the sum in (5.1.11) vanishes in this case, as desired. All possibilities of $\mathcal{A} \not\leq \mathcal{E}$ are now covered. In all cases (5.1.11) holds.

The arguments for showing that the sum in (5.1.12) vanishes if $\mathcal{A} \not\leq \mathcal{E}$ are similar.

$\square$

5.2. A correspondence for lattice paths that cross the line $x = y$.

We wish to handle the strange major counting of paths from $\mathcal{A} = (A_1, A_2)$ to $\mathcal{E} = (E_1, E_2)$ which do not cross the line $x = y$. Recall that ordinary counting of these paths is done by the reflection principle (see the second part of subsection 2.2). Suppose that $A_1 \geq A_2$ and $E_1 \geq E_2$, i.e. that $\mathcal{A}$ and $\mathcal{E}$ lie below or maybe on the line $x = y$. Remember that when saying that a path "lies below $x = y$" we allow the path to touch $x = y$. The reflection principle shows that the paths from $\mathcal{A}$ to $\mathcal{E}$ which *cross* $x = y$ are in one-to-one correspondence with the paths from $(A_2 - 1, A_1 + 1)$ to (E_1, E_2).

To undertake the strange major counting, the reflection has to be replaced by a bijection having similar properties, because we cannot use the reflection itself since, when reflecting a part of a path, nothing can be said about the major index or the strange major index of its image. We construct two fundamental bijections for crossing paths, one for the case $A_1 \geq A_2$, $E_1 \geq E_2$, and one for the case $A_1 \leq A_2$, $E_1 \leq E_2$. The first bijection is the analogue for the reflection procedure described in subsection 2.2. It maps paths P from (A_1, A_2) to (E_1, E_2) which cross $x = y$ onto paths P' from $(A_2 - 1, A_1 + 1)$ to (E_1, E_2) in a way that only the initial portion till the last point lying above $x = y$ is changed. These properties are shared with the reflection procedure. In addition, for this mapping there holds a relation between $\mathrm{ymaj}_{\beta;\gamma}(P)$ and $\mathrm{xmaj}_{\beta;\gamma+1}(P')$.

Proposition 24. FIRST FUNDAMENTAL BIJECTION FOR CROSSING PATHS. *Let* $\mathcal{A} = (A_1, A_2)$ *and* $\mathcal{E} = (E_1, E_2)$ *be two lattice points in the integer lattice* $\mathbb{Z}^2$ *with* $A_1 \geq A_2$ *and* $E_1 \geq E_2$. *For every integer* γ *satisfying*

$$A_2 \leq \gamma \leq A_1 \tag{5.2.1}$$

there is a bijection $\Phi_\gamma^{(1)}$ *between paths* P *from* $\mathcal{A}$ *to* $\mathcal{E}$ *which cross the line* $x = y$ *and the paths from* $(A_2 - 1, A_1 + 1)$ *to* (E_1, E_2), *such that the last meeting point of* P *with* $x = y - 1$ *is identical with the last meeting point of* $\Phi_\gamma^{(1)}(P)$ *with* $x = y - 1$, *such that the path's portions after this meeting point are identical, and*

$$\mathrm{ymaj}_{\beta;\gamma}(P) = \mathrm{xmaj}_{\beta;\gamma+1}(\Phi_\gamma^{(1)}(P)) + A_1 - A_2 + 1 + \beta. \tag{5.2.2}$$

REMARK. The relevance of the assumption $\gamma \leq A_1$ lies in the fact that it allows a path from $\mathcal{A}$ to $\mathcal{E}$ to cross $x = y$ only before having crossed the horizontal line $y = \gamma$. After having crossed $x = y$ for the first time, the path must have at least one North-East corner before returning to $x = y$. This North-East corner, because of lying above $y = \gamma$, contributes a β to the statistics $\mathrm{ymaj}_{\beta;\gamma}$. This explains the β in (5.2.2).

PROOF OF THE PROPOSITION. First we consider the case $\gamma \geq E_2$. It is clear that for each point (p_1, p_2) of a path P from $\mathcal{A}$ to $\mathcal{E}$ we then have $p_1 \geq A_1 \geq \gamma \geq E_2 \geq p_2$. This simply means that P cannot cross the line $x = y$ but stays below everywhere only being allowed to touch it. Therefore the set of crossing paths is empty as well as the set of paths from $(A_2 - 1, A_1 + 1)$ to (E_1, E_2) because of $A_1 \geq E_2$. This implies that the Proposition in this case trivially holds.

Now let $\gamma < E_2$. The construction of $\Phi_\gamma^{(1)}$ is done in four steps. For the reader who is not interested in all the details, we outline a short-cut through the following construction. The mapping is essentially explained by the lines (5.2.3a,b), (5.2.4), (5.2.10)–(5.2.15), and the first passage of the fourth step. An example is added at the end of the proof.

Suppose that P is a path from $\mathcal{A}$ to $\mathcal{E}$ which crosses $x = y$.

First step. We map P onto its array representation (cf. (2.1.2)). This gives a bijection between crossing paths P and arrays $(a \mid b) = (a_1, \ldots, a_k \mid b_1, \ldots, b_k)$,

$$\text{paths } P : (A_1, A_2) \to (E_1, E_2) \text{ which cross } x = y \qquad (5.2.3a)$$

$$\updownarrow$$

$$\begin{matrix} A_1 \leq & a_1 & \cdots & a_k & \leq E_1 - 1 \\ A_2 + 1 \leq & b_1 & \cdots & b_k & \leq E_2 \end{matrix}, \text{ there exists an } i, 1 \leq i \leq k \text{ with } a_i < b_i \qquad (5.2.3b)$$

which (because of (4.0.6)) satisfies

$$\mathrm{ymaj}_{\beta;\gamma}(P) = \|a\| + \|b\| + \beta \cdot |\{t : b_t > \gamma\}| - k(A_1 + A_2). \qquad (5.2.3c)$$

The condition required for the arrays in (5.2.3b) is clear from the construction of the array (which was described in subsection 2.1) and the fact that a path's North-East corner (a_i, b_i) which lies above $x = y$ satisfies $a_i < b_i$.

Second step. Here we use a modification of a correspondence first introduced in [26, section 4]. Given an array $(a \mid b)$ of the type (5.2.3b) let I be maximal with $a_I < b_I$. We map $(a \mid b)$ onto another two-rowed array by

$$\begin{matrix} a_1 & \cdots\cdots\cdots & a_I & a_{I+1} & \cdots & a_k \\ b_1 & \cdots & b_{I-1} & b_I & \cdots\cdots\cdots & b_k \end{matrix}$$

$$\longrightarrow \quad \begin{matrix} & b_1 & \cdots\cdots & b_{I-1} & a_{I+1} & \cdots & a_k \\ a_1\, a_2\, a_3 & \cdots & a_I & b_I & \cdots\cdots\cdots & b_k \end{matrix}. \qquad (5.2.4)$$

(The vertical and horizontal lines in (5.2.4) are drawn only to support the orientation of the reader. They visually separate the parts of the arrays that are changed from the parts that remain unchanged.) More precisely, the new array $(\bar{a} \mid \bar{b}) = (\bar{a}_2, \ldots, \bar{a}_k \mid \bar{b}_0, \ldots, \bar{b}_k)$ is given by

$$\bar{a}_i = \begin{cases} b_{i-1} & 2 \le i \le I \\ a_i & I < i \le k \end{cases}$$

and

$$\bar{b}_i = \begin{cases} a_{i+1} & 0 \le i < I \\ b_i & I \le i \le k \end{cases}.$$

Obviously the new array $(\bar{a} \mid \bar{b})$ is strictly increasing in both rows. Moreover, I is maximal with $\bar{a}_I < \bar{b}_I$. Therefore the inverse of the mapping (5.2.4) simply is

$$
\begin{array}{ccccccc}
& \bar{a}_2 & \cdots\cdots & \bar{a}_I & \bar{a}_{I+1} & \cdots & \bar{a}_k \\
\bar{b}_0\ \bar{b}_1\ \bar{b}_2 & \cdots & \bar{b}_{I-1} & \bar{b}_I & \cdots\cdots\cdots & & \bar{b}_k
\end{array}
$$

$$
\longrightarrow
\begin{array}{ccccccc}
\bar{b}_0 & \cdots\cdots & \bar{b}_{I-1} & \bar{a}_{I+1} & \cdots & \bar{a}_k \\
\bar{a}_2 & \cdots & \bar{a}_I & \bar{b}_I & \cdots\cdots\cdots & \bar{b}_k
\end{array},
$$

where I is maximal with $\bar{a}_I < \bar{b}_I$. If there is no such I then $I = 1$. From this it follows that the mapping (5.2.4) is a bijection between the following two families of arrays

$$
\begin{array}{ccccc}
A_1 \le & a_1 & \cdots & a_k & \le E_1 - 1 \\
A_2 + 1 \le & b_1 & \cdots & b_k & \le E_2
\end{array}
\text{, there is an } i \text{ with } a_i < b_i \tag{5.2.5a}
$$

$$
\longleftrightarrow
\begin{array}{cccccc}
A_2 + 1 \le & & \bar{a}_2 & \cdots & \bar{a}_k & \le E_1 - 1 \\
A_1 \le & \bar{b}_0 & \bar{b}_1\ \bar{b}_2 & \cdots & \bar{b}_k & \le E_2
\end{array}
\tag{5.2.5b}
$$

which satisfies

$$\|a\| + \|b\| + \beta \cdot |\{t : b_t > \gamma\}| = \|\bar{a}\| + \|\bar{b}\| + \beta \cdot |\{t : \bar{a}_t > \gamma\}| + \beta. \tag{5.2.5c}$$

The last identity is due to the assumption $\gamma \le A_1$. Namely, for an array $(a \mid b)$ of the type (5.2.3b) and I being maximal with $a_I < b_I$, we have $\gamma \le A_1 \le a_I < b_I$. Therefore whenever $b_i \le \gamma$, there holds $i < I$. Let L be maximal with $b_L \le \gamma$ and let $(\bar{a} \mid \bar{b})$ the array which by (5.2.4) corresponds to $(a \mid b)$. Then by means of (5.2.4) there holds

$$|\{t : b_t > \gamma\}| = k - L$$

and

$$
\begin{aligned}
|\{t : \bar{a}_t > \gamma\}| &= |\{t : \bar{a}_t > \gamma \text{ and } 2 \le t \le I\}| \\
&\quad + |\{t : \bar{a}_t > \gamma \text{ and } I < t \le k\}| \\
&= |\{t : b_{t-1} > \gamma \text{ and } 2 \le t \le I\}| \\
&\quad + |\{t : a_t > \gamma \text{ and } I < t \le k\}| \\
&= (I - L - 1) + (k - I) \\
&= k - L - 1,
\end{aligned}
$$

which settles (5.2.5c).

An auxiliary correspondence. Before being able to perform the next step we need an auxiliary correspondence. We claim that there is a bijection between the following two families of arrays,

$$
\begin{array}{ccccccc}
1 \leq & & & c_2 & \ldots & c_{I-1} & c_I & \leq M-1 \\
0 \leq & d_0 & d_1 & d_2 & \ldots & d_{I-1} & & \leq N-1
\end{array}
\tag{5.2.6a}
$$

$$
\longleftrightarrow
\begin{array}{ccccc}
0 \leq & \bar{c}_1 & \ldots & \bar{c}_{J-1} & \bar{c}_J & \leq M \\
1 \leq & \bar{d}_1 & \ldots & \bar{d}_{J-1} & & \leq N-2
\end{array}
\tag{5.2.6b}
$$

such that

$$
\|c\| + \|d\| + \beta \cdot |\{t : c_t > \overline{\gamma}\}| = \|\bar{c}\| + \|\bar{d}\| + \beta \cdot |\{t : \bar{c}_t > \overline{\gamma} + 1\}|
\tag{5.2.6c}
$$

for all integers $\overline{\gamma}$ satisfying $0 \leq \overline{\gamma} \leq M-1$. The easiest way to see this, is to show that the corresponding generating functions are equal. For arrays of the type (5.2.6a) we obtain

$$
\sum_{\substack{I \geq 1 \\ c \in P_{I-1}[1, M-1] \\ d \in P_I[0, N-1]}} q^{\|c\| + \|d\| + \beta \cdot |\{t : c_t > \overline{\gamma}\}|}
\tag{5.2.7a}
$$

$$
= \sum_{I \geq 1} \left(\sum_{d \in P_I[0, N-1]} q^{\|d\|} \right)
$$

$$
\times \sum_{i+j=I-1} \left(\sum_{c' \in P_i[1, \overline{\gamma}]} q^{\|c'\|} \right) \left(\sum_{c'' \in P_j[\overline{\gamma}+1, M-1]} q^{\|c''\| + \beta \cdot |\{t : c''_t > \overline{\gamma}\}|} \right)
$$

$$
= \sum_{I \geq 1} q^{\binom{I}{2}} \begin{bmatrix} N \\ I \end{bmatrix} \sum_{i+j=I-1} q^{\binom{i+1}{2}} \begin{bmatrix} \overline{\gamma} \\ i \end{bmatrix} q^{j(\beta+\overline{\gamma}) + \binom{j+1}{2}} \begin{bmatrix} M - \overline{\gamma} - 1 \\ j \end{bmatrix}
$$

$$
= \sum_{j \geq 0} q^{j(j+\beta+\overline{\gamma}+1)} \begin{bmatrix} M - \overline{\gamma} - 1 \\ j \end{bmatrix} \sum_{I \geq 1} q^{I(I-1-j)} \begin{bmatrix} N \\ I \end{bmatrix} \begin{bmatrix} \overline{\gamma} \\ I-1-j \end{bmatrix}
\tag{5.2.7b}
$$

$$
= \sum_{j \geq 0} q^{j(j+\beta+\overline{\gamma}+1)} \begin{bmatrix} M - \overline{\gamma} - 1 \\ j \end{bmatrix} \begin{bmatrix} N + \overline{\gamma} \\ N-1-j \end{bmatrix},
\tag{5.2.7c}
$$

where in the last step we used the q-Vandermonde summation (5.1.4). On the other hand for arrays of the type (5.2.6b) we get

$$\sum_{\substack{J \geq 1}} \sum_{\substack{\bar{c} \in P_J[0,M] \\ \bar{d} \in P_{J-1}[1,N-2]}} q^{\|\bar{c}\| + \|\bar{d}\| + \beta \cdot |\{t : \bar{c}_t > \bar{\gamma}+1\}|} \tag{5.2.8a}$$

$$= \sum_{J \geq 1} \left(\sum_{d \in P_{J-1}[1,N-2]} q^{\|d\|} \right)$$

$$\times \sum_{i+j=J} \left(\sum_{c' \in P_i[0,\bar{\gamma}+1]} q^{\|c'\|} \right) \left(\sum_{c'' \in P_j[\bar{\gamma}+2,M]} q^{\|c''\| + \beta \cdot |\{t : c''_t > \bar{\gamma}+1\}|} \right)$$

$$= \sum_{J \geq 1} q^{\binom{J}{2}} \begin{bmatrix} N-2 \\ J-1 \end{bmatrix} \sum_{i+j=J} q^{\binom{i}{2}} \begin{bmatrix} \bar{\gamma}+2 \\ i \end{bmatrix} q^{j(\beta+\bar{\gamma}+2)+\binom{j}{2}} \begin{bmatrix} M-\bar{\gamma}-1 \\ j \end{bmatrix}$$

$$= \sum_{j \geq 0} q^{j(j+\beta+\bar{\gamma}+1)} \begin{bmatrix} M-\bar{\gamma}-1 \\ j \end{bmatrix} \sum_{J \geq 1} q^{(J-1)(J-j)} \begin{bmatrix} N-2 \\ J-1 \end{bmatrix} \begin{bmatrix} \bar{\gamma}+2 \\ J-j \end{bmatrix} \tag{5.2.8b}$$

$$= \sum_{j \geq 0} q^{j(j+\beta+\bar{\gamma}+1)} \begin{bmatrix} M-\bar{\gamma}-1 \\ j \end{bmatrix} \begin{bmatrix} N+\bar{\gamma} \\ N-1-j \end{bmatrix} , \tag{5.2.8c}$$

by renewed application of the q-Vandermonde summation. Since (5.2.7c) and (5.2.8c) obviously are identical, the generating functions (5.2.7a) and (5.2.8a) agree. Hence, there must be a bijection satisfying (5.2.6).

This is only an existence proof which does not please any true combinatorialist. But the computations (5.2.7) and (5.2.8) tell us how to explicitly construct such a bijection. For sake of completeness we give a description of it below. We leave it to the reader to verify that it is well-defined and satisfies the desired properties (5.2.6).

Let $(c \mid d)$ be an array of the type (5.2.6a). To illustrate the following we choose a running example, say $M = 11$, $N = 10$, $\bar{\gamma} = 7$, and the array $(c_{ex} \mid d_{ex})$ given by

$$\begin{array}{ccc} 1 \leq & 1\ 4\ 5\ 7\ 8 & \leq 10 \\ 0 \leq & 1\ 3\ 6\ 7\ 8\ 9 & \leq 9 \end{array},$$

or more precisely $c_{ex} = \{1, 4, 5, 7, 8\}$ and $d_{ex} = \{1, 3, 6, 7, 8, 9\}$.

First c is split up into the initial part c' and the final part c'', where c' contains all elements of c which are smaller or equal $\bar{\gamma}$, and c'' contains the rest. In c' add N to each element to obtain c'''. In our example we have $c'_{ex} = \{1, 4, 5, 7\}$, $c''_{ex} = \{8\}$, and $c'''_{ex} = \{11, 14, 15, 17\}$. Let d' denote the complement of d with respect to $\{0, 1, \ldots, N-1\}$. In d' replace each element e by $N-e$, thus obtaining d''. In our example we have $d'_{ex} = \{0, 2, 4, 5\}$ and $d''_{ex} = \{5, 6, 8, 10\}$. Now form the union of d'' and c''' and denote it by u. In our example this is $\{5, 6, 8, 10, 11, 14, 15, 17\}$.

Next u is split up into $\bar{d}''$ and $\bar{c}'''$ where $\bar{d}''$ contains all elements of u being smaller or equal $\bar{\gamma} + 1$ and $\bar{c}'''$ the rest. In our example we have $\bar{d}''_{ex} = \{5, 6, 8\}$ and $\bar{c}'''_{ex} = \{10, 11, 14, 15, 17\}$. Now in $\bar{d}''$ we replace each element e by $N-1-e$, thus obtaining $\bar{d}'$. The complement of $\bar{d}'$ with respect to $\{1, 2, \ldots, N-2\}$ is denoted by $\bar{d}$. On the other hand we subtract $N-1$ from each element of $\bar{c}'''$ and thus get $\bar{c}'$. The set $\bar{c}''$

is formed by adding 1 to each element of c''. In our example we get $\bar{d}'_{\text{ex}} = \{1,3,4\}$, $\bar{d}_{\text{ex}} = \{2,5,6,7,8\}$, $\bar{c}'_{\text{ex}} = \{1,2,5,6,8\}$, and $\bar{c}''_{\text{ex}} = \{9\}$. Finally we arrive at the array $(\bar{c} \mid \bar{d})$ by defining $\bar{c}$ to be the union of $\bar{c}'$ and $\bar{c}''$.

In our example this gives the correspondence

$$
\begin{array}{llll}
1 \leq & 1\,4\,5\,7\,8 & \leq 10 \\
0 \leq & 1\,3\,6\,7\,8\,9 & \leq \ 9
\end{array}
\quad \longleftrightarrow \quad
\begin{array}{lll}
0 \leq & 1\,2\,5\,6\,8\,9 & \leq 11 \\
1 \leq & 2\,5\,6\,7\,8 & \leq \ 8
\end{array}
\tag{5.2.9}
$$

It is a routine verification that this above defined mapping is a bijection between the sets of arrays in (5.2.6a,b) which satisfies (5.2.6c). As mentioned above, the motivation of this construction lies in the computations (5.2.7) and (5.2.8). What we did was to interpret each step in these computations combinatorially. Starting at (5.2.7a) first c is split up into c' and c''. Then for the moment we forget about c'' and consider only d and c' which corresponds to the reordering of summations in (5.2.7b). In order to have a combinatorial version of the next step, the q-Vandermonde summation, we first had to transform d into d''. Roughly spoken this corresponds to transforming $\begin{bmatrix} N \\ I \end{bmatrix}$ to $\begin{bmatrix} N \\ N-I \end{bmatrix}$ in (5.2.7b). Now we merged d'' and c''' (which has been a shifted version of c') together, what could be called the combinatorial q-Vandermonde summation. Now we arrived at (5.2.7c) and therefore are able to continue at (5.2.8c). It should be clear that splitting up u, etc., has been nothing but working backward from (5.2.8c) to (5.2.8a).

Third step. Recall that by the first and second step we mapped a path P from $\mathcal{A}$ to $\mathcal{E}$ which crosses $x = y$ to an array of the type (5.2.5b). Besides, we denoted by I the maximal index for which $\bar{a}_I < \bar{b}_I$. Given an array of the type (5.2.5b) and its correlated index I,

$$
\begin{array}{llllllll}
A_2 + 1 \leq & & \bar{a}_2 & \cdots\cdots\cdots & \bar{a}_I \mid \bar{a}_{I+1} & \cdots & \bar{a}_k & \leq E_1 - 1 \\
A_1 \leq & \bar{b}_0\ \bar{b}_1\ \bar{b}_2 & \cdots & \bar{b}_{I-1} \mid \bar{b}_I & \cdots\cdots\cdots & & \bar{b}_k & \leq E_2
\end{array}
\tag{5.2.10}
$$

for the moment we only consider the subarray

$$
\begin{array}{llll}
& & \bar{a}_2\ \cdots\ \bar{a}_{I-1}\ \bar{a}_I \\
\bar{b}_0\ \bar{b}_1\ \bar{b}_2 & \cdots & \bar{b}_{I-1}
\end{array}
\quad,
\tag{5.2.11}
$$

or with bounds included (note the upper bounds!)

$$
\begin{array}{lll}
A_2 + 1 \leq & \bar{a}_2\ \cdots\ \bar{a}_{I-1}\ \bar{a}_I & \leq \bar{b}_I - 1 \\
A_1 \leq & \bar{b}_0\ \bar{b}_1\ \bar{b}_2\ \cdots\ \bar{b}_{I-1} & \leq \bar{b}_I - 1
\end{array}
\quad .
$$

(Similar to (5.2.4), the vertical and horizontal lines in (5.2.10) are drawn only to help the reader in separating the subarray from the rest.) We subtract A_2 from each entry of the first row and A_1 from each entry of the second row of this subarray. It is clear that the resulting array is of the type (5.2.6a) with the choices $M = \bar{b}_I - A_2$, $N = \bar{b}_I - A_1$, $\bar{\gamma} = \gamma - A_2$,

$$
\begin{array}{llllll}
1 \leq & & \bar{a}_2 - A_2 & \cdots & \bar{a}_{I-1} - A_2\ \ \bar{a}_I - A_2 & \leq \bar{b}_I - A_2 - 1 \\
0 \leq & \bar{b}_0 - A_1\ \ \bar{b}_1 - A_1\ \ \bar{b}_2 - A_1 & \cdots & \bar{b}_{I-1} - A_1 & \leq \bar{b}_I - A_1 - 1
\end{array}
\quad .
\tag{5.2.12}
$$

Because of the assumption $\gamma \geq A_2$ and because of $\gamma < b_I$ and $b_I = \bar{b}_I$ (this results from the second step) the difference $\gamma - A_2$ indeed satisfies $0 \leq \gamma - A_2 \leq \bar{b}_I - A_2 - 1$. Therefore by (5.2.6a,b) the array (5.2.12) uniquely corresponds to an array of the type

$$
\begin{array}{ccccc}
0 \leq & \tilde{a}_1 & \ldots & \tilde{a}_{J-1} & \tilde{a}_J & \leq \bar{b}_I - A_2 \\
1 \leq & \tilde{b}_1 & \ldots & \tilde{b}_{J-1} & & \leq \bar{b}_I - A_1 - 2
\end{array}
\tag{5.2.13}
$$

Next we add $A_2 - 1$ to each entry in the first row and $A_1 + 1$ to each entry in the second row to obtain the array

$$
\begin{array}{ccccc}
A_2 - 1 \leq & \tilde{a}_1 + A_2 - 1 & \ldots & \tilde{a}_{J-1} + A_2 - 1 & \tilde{a}_J + A_2 - 1 & \leq \bar{b}_I - 1 \\
A_1 + 2 \leq & \tilde{b}_1 + A_1 + 1 & \ldots & \tilde{b}_{J-1} + A_1 + 1 & & \leq \bar{b}_I - 1
\end{array}
\tag{5.2.14}
$$

Finally this array can be put together with the rest of (5.2.10),

$$
\begin{array}{ccc}
A_2 - 1 \leq & \tilde{a}_1 + A_2 - 1 \ldots \ldots \ldots \ldots \ldots \tilde{a}_J + A_2 - 1 \mid \bar{a}_{I+1} \ldots \bar{a}_k & \leq E_1 - 1 \\
A_1 + 2 \leq & \tilde{b}_1 + A_1 + 1 \ldots \tilde{b}_{J-1} + A_1 + 1 \mid \quad \bar{b}_I \qquad \ldots \ldots \ldots b_k & \leq E_2
\end{array},
\tag{5.2.15}
$$

thus obtaining an array of the type

$$
\begin{array}{ccccc}
A_2 - 1 \leq & \hat{a}_1 & \ldots & \hat{a}_m & \leq E_1 - 1 \\
A_1 + 2 \leq & \hat{b}_1 & \ldots & \hat{b}_m & \leq E_2
\end{array}.
\tag{5.2.16}
$$

How are the weights of (5.2.10) and (5.2.16) related? Assuming that $(\hat{a} \mid \hat{b})$ is defined by (5.2.15), by means of (5.2.6c) and (5.2.10)–(5.2.15) we have

$$
\|\bar{a}\| + \|\bar{b}\| + \beta \cdot |\{t : \bar{a}_t > \gamma\}|
$$
$$
= \|\hat{a}\| + \|\hat{b}\| + \beta \cdot |\{t : \hat{a}_t \geq \gamma + 1\}|
$$
$$
+ (I - 1)A_2 + IA_1 - J(A_2 - 1) - (J - 1)(A_1 + 1)
$$
$$
= \|\hat{a}\| + \|\hat{b}\| + \beta \cdot |\{t : \hat{a}_t \geq \gamma + 1\}| + (I - J)(A_1 + A_2) + A_1 - A_2 + 1.
$$

Subtraction of $k(A_1 + A_2)$ then gives

$$
\|\bar{a}\| + \|\bar{b}\| + \beta \cdot |\{t : \bar{a}_t > \gamma\}| - k(A_1 + A_2)
$$
$$
= \|\hat{a}\| + \|\hat{b}\| + \beta \cdot |\{t : \hat{a}_t \geq \gamma + 1\}| - (k - I + J)(A_1 + A_2) + A_1 - A_2 + 1.
\tag{5.2.17}
$$

Fourth step. Obviously, arrays of the type (5.2.16) are in one-to-one correspondence with paths $\hat{P}$ from $(A_2 - 1, A_1 + 1)$ to (E_1, E_2), and besides there holds

$$
\text{xmaj}_{\beta;\gamma+1}(\hat{P}) = \|\hat{a}\| + \|\hat{b}\| + \beta \cdot |\{t : \hat{a}_t \geq \gamma + 1\}| - m(A_2 - 1 + A_1 + 1). \tag{5.2.18}
$$

Now we are able to define $\Phi_\gamma^{(1)}$. For any path P from $\mathcal{A}$ to $\mathcal{E}$ which crosses $x = y$ perform the steps 1–4. The image of P under application of $\Phi_\gamma^{(1)}$ by definition is the resulting path $\hat{P}$,

$$
\Phi_\gamma^{(1)}(P) := \hat{P}.
$$

Combining (5.2.3c), (5.2.5c), (5.2.17), and (5.2.18), we obtain the desired property (5.2.2). Here it is important to notice that the row lengths of the array in (5.2.15) equal $k - I + J$.

Since all steps which led to the construction of $\Phi_\gamma^{(1)}$ of a crossing path P are invertible, $\Phi_\gamma^{(1)}$ is a bijection.

The only thing which is still open is the property about the last meeting point with the line $x = y - 1$. Recall that in the second step I was maximal with $a_I < b_I$. Having in mind the connection between a path P and its corresponding array representation $(a \mid b)$, we see that the last meeting point of P with $x = y - 1$ is $(b_I - 1, b_I)$. Obviously, the construction of $\Phi_\gamma^{(1)}(P)$ has the property that the last $k - I$ elements in the first row of the array in (5.2.3b) (i.e. the elements $a_{I+1}, \ldots, a_k$) are identical with the last $k - I$ elements in the first row of the array in (5.2.15), the same being valid for the last $k - I + 1$ elements in the second rows (i.e. the elements $b_I, \ldots, b_k$). Besides by construction there holds $\tilde{a}_J + A_2 - 1 < \bar{b}_I = b_I$ in (5.2.15). Hence, the last meeting point of $\Phi_\gamma^{(1)}(P)$ with $x = y - 1$ is $(b_I - 1, b_I)$, too, and the path portions of P and $\Phi_\gamma^{(1)}(P)$ after this last meeting point are identical.

This completes the proof of the Proposition.

To illustrate the action of $\Phi_\gamma^{(1)}$, consider the path $P_{\text{ex}} = ((-1, -2),$ 12112221112122121222211211112212) with starting point $\mathcal{A}_{\text{ex}} = (-1, -2)$ and final point $\mathcal{E}_{\text{ex}} = (15, 13)$. It is the lower path in Figure 4. Choose $\gamma = 5$. In the first step P_{ex} is mapped to its array representation $(a^{(\text{ex})} \mid b^{(\text{ex})})$ given by

$$
\begin{array}{ccc}
-1 \leq & 0\ 2\ 5\ 6\ 7\ 8\ 10\ 14 & \leq 14 \\
-1 \leq & -1\ 2\ 3\ 5\ 6\ 9\ 10\ 12 & \leq 13
\end{array} .
$$

The index I which is maximal with respect to $a_I^{(\text{ex})} < b_I^{(\text{ex})}$ is $I = 6$. Hence, in the second step by (5.2.4) the array $(a^{(\text{ex})} \mid b^{(\text{ex})})$ is mapped to the array

$$
\begin{array}{ccc}
-1 \leq & -1\ 2\ 3\ 5\ 6\ 10\ 14 & \leq 14 \\
-1 \leq & 0\ 2\ \ 5\ 6\ 7\ 8\ 9\ 10\ 12 & \leq 13
\end{array} . \tag{5.2.19}
$$

To perform the third step, we first have to restrict our attention to the subarray

$$
\begin{array}{ccc}
-1 \leq & -1\ 2\ 3\ 5\ 6 & \leq 8 \\
-1 \leq & 0\ 2\ \ 5\ 6\ 7\ 8 & \leq 8
\end{array} .
$$

By virtue of (5.2.12) we obtain the array

$$
\begin{array}{ccc}
1 \leq & 1\ 4\ 5\ 7\ 8 & \leq 10 \\
0 \leq & 1\ 3\ 6\ 7\ 8\ 9 & \leq 9
\end{array} .
$$

Now we have to apply the auxiliary correspondence to this array. Luckily it just agrees with the left-hand array of (5.2.9) and therefore corresponds to

$$
\begin{array}{ccc}
0 \leq & 1\ 2\ 5\ 6\ 8\ 9 & \leq 11 \\
1 \leq & 2\ 5\ 6\ 7\ 8 & \leq 8
\end{array} ,
$$

the right-hand array in (5.2.9). By (5.2.14), from this array we obtain

$$
\begin{array}{ccc}
-3 \leq & -2\ -1\ 2\ 3\ 5\ 6 & \leq 8 \\
1 \leq & 2\ \ 5\ 6\ 7\ 8 & \leq 8
\end{array} .
$$

By (5.2.15), this is glued together with the rest of (5.2.19) to obtain

$$
\begin{array}{rl}
-3 \leq & -2\ -1\ 2\ 3\ 5\ 6\ 10\ 14\ \ \leq 14 \\
1 \leq & 2\ \ \ 5\ 6\ 7\ 8\ 9\ 10\ 12\ \ \leq 13
\end{array}
$$

which is the array representation of the second path in Figure 4. The last meeting point of P_{ex} with $x = y - 1$ is $\mathcal{S}_{\mathrm{ex}} = (8, 9)$, it is the same as that of $\Phi_5^{(1)}(P_{\mathrm{ex}})$. Besides these two paths obviously are identical after $\mathcal{S}_{\mathrm{ex}}$, as was claimed.

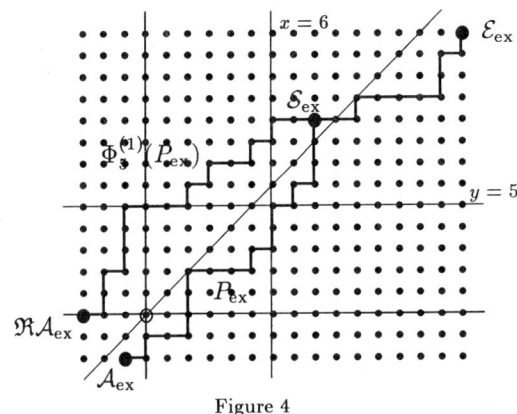

Figure 4

By definition (4.0.1), we have

$$
\mathrm{ymaj}_{\beta;5}(P_{\mathrm{ex}}) = 2 + 7 + 11 + 14 + 16 + 20 + 23 + 29 + 4\beta = 122 + 4\beta,
$$

and by (4.0.2) we have

$$
\mathrm{xmaj}_{\beta;6}(\Phi_5^{(1)}(P_{\mathrm{ex}})) = 3 + 7 + 11 + 13 + 16 + 18 + 23 + 29 + 3\beta = 120 + 3\beta,
$$

in accordance with (5.2.2).   □

Setting $\beta = 0$ in Proposition 24, we obtain as a corollary a bijection for the ordinary major index.

Corollary 25. *Let $\mathcal{A} = (A_1, A_2)$ and $\mathcal{E} = (E_1, E_2)$ be two lattice points in the integer lattice $\mathbb{Z}^2$ with $A_1 \geq A_2$ and $E_1 \geq E_2$. There is a bijection $\Phi^{(1)}$ between paths P from $\mathcal{A}$ to $\mathcal{E}$ which cross the line $x = y$ and the paths from $(A_2 - 1, A_1 + 1)$ to (E_1, E_2) such that the last meeting point of P with $x = y - 1$ is identical with the last meeting point of $\Phi^{(1)}(P)$ with $x = y - 1$, such that the path's portions after this meeting point are identical, and*

$$
\mathrm{maj}\, P = \mathrm{maj}\, \Phi^{(1)}(P) + A_1 - A_2 + 1. \tag{5.2.20}
$$

PROOF. Set $\beta = 0$ and define $\Phi^{(1)} := \Phi_{A_2}^{(1)}$.   □

A simpler way to define a bijection satisfying the properties of the above Corollary has been previously given in [27, Proof of (5.36)] by extending an idea of Fürlinger and Hofbauer [13, Lemma on p. 255]. But this bijection is only applicable for the major index and is not extendable in order to be applicable with the strange major index.

The next Proposition provides an analogue of Proposition 24 in the case that the starting point and the final point of the paths both lie above the line $x = y$.

Proposition 26. SECOND FUNDAMENTAL BIJECTION FOR CROSSING PATHS. *Let $\mathcal{A} = (A_1, A_2)$ and $\mathcal{E} = (E_1, E_2)$ be two lattice points in the integer lattice $\mathbb{Z}^2$ with $A_1 \leq A_2$ and $E_1 \leq E_2$. There is a bijection $\Phi^{(2)}$ between paths P from $\mathcal{A}$ to $\mathcal{E}$ which cross the line $x = y$ and the paths from $(A_2 + 1, A_1 - 1)$ to (E_1, E_2), such that the last meeting point of P with $x = y + 1$ is identical with the last meeting point of $\Phi^{(2)}(P)$ with $x = y + 1$, such that the path's portions after this meeting point are identical, and which satisfies*

$$\mathrm{xmaj}_{\beta;\gamma}(P) = \mathrm{ymaj}_{\beta;\gamma-1}(\Phi^{(2)}(P)) \qquad (5.2.21)$$

for every integer γ with $A_1 \leq \gamma \leq A_2$, and also

$$\mathrm{maj}\, P = \mathrm{maj}\, \Phi^{(2)}(P). \qquad (5.2.22)$$

REMARK. Quite similarly to the Remark after Proposition 24, it should be observed that the relevance of the assumption $\gamma \leq A_2$ lies in the fact that this allows a path from $\mathcal{A}$ to $\mathcal{E}$ to cross $x = y$ only after before having crossed the vertical line $x = \gamma$. Again this will be important in the proof.

PROOF OF THE PROPOSITION. The procedure is quite similar to that in the proof of Proposition 24. It is even simpler since this auxiliary correspondence and therefore also the third step can be avoided. So it will not be necessary to go into all details, a description without comments should suffice. For a better understanding of the following the reader might look up the example at the end of the proof.

First we consider the case $\gamma \geq E_1$. It is clear that for each point (p_1, p_2) of a path P from $\mathcal{A}$ to $\mathcal{E}$ we then have $p_2 \geq A_2 \geq \gamma \geq E_1 \geq p_1$. This simply means that P cannot cross the line $x = y$ but stays below everywhere only being allowed to touch it. Therefore the set of crossing paths is empty as well as the set of paths from $(A_2 + 1, A_1 - 1)$ to (E_1, E_2) because of $A_2 \geq E_1$. This implies that the Proposition in this case trivially holds.

Now let $\gamma < E_1$. The construction of $\Phi^{(2)}$ this time is done in three steps. Suppose that P is a path from $\mathcal{A}$ to $\mathcal{E}$ which crosses $x = y$.

First step. Use the array representation (2.1.2) of paths to see that this provides a bijection between the following sets,

$$\text{paths } P : (A_1, A_2) \to (E_1, E_2) \text{ which cross } x = y \qquad (5.2.23a)$$

$$\updownarrow$$

$$\begin{matrix} A_1 \leq & a_1 & \ldots & a_k & \leq E_1 - 1 \\ A_2 + 1 \leq & b_1 & \ldots & b_k & \leq E_2 \end{matrix}, \text{ there exists an } i, 1 \leq i \leq k - 1 \text{ with } a_{i+1} > b_i$$
$$(5.2.23b)$$

which (because of (4.0.7)) satisfies

$$\mathrm{xmaj}_{\beta;\gamma}(P) = \|a\| + \|b\| + \beta \cdot |\{t : a_t \geq \gamma\}| - k(A_1 + A_2). \tag{5.2.23c}$$

Second step. For an array $(a \mid b)$ of the type (5.2.23b) let I be maximal with $a_{I+1} > b_I$. We map $(a \mid b)$ to another two-rowed array $(\bar{a} \mid \bar{b})$ by

$$\begin{array}{ccccccc}
A_1 \leq & a_1 & \ldots & a_I & a_{I+1} & \ldots & a_k & \leq E_1 - 1 \\
A_2 + 1 \leq & b_1 & \ldots & b_I & b_{I+1} & \ldots & b_k & \leq E_2
\end{array} \tag{5.2.24a}$$

$$\longleftrightarrow \quad \begin{array}{ccccccc}
A_2 + 1 \leq & b_1 & \ldots & b_I & a_{I+1} & \ldots & a_k & \leq E_1 - 1 \\
A_1 \leq & a_1 & \ldots & a_I & b_{I+1} & \ldots & b_k & \leq E_2
\end{array} \tag{5.2.24b}$$

(Again the vertival lines in (5.2.24) are drawn only to show which parts of the arrays are changed and which not.) More precisely, the new array $(\bar{a} \mid \bar{b}) = (\bar{a}_1, \ldots, \bar{a}_k \mid \bar{b}_1, \ldots, \bar{b}_k)$ is given by

$$\bar{a}_i = \begin{cases} b_i & 1 \leq i \leq I \\ a_i & I < i \leq k \end{cases}$$

and

$$\bar{b}_i = \begin{cases} a_i & 1 \leq i \leq I \\ b_i & I < i \leq k \end{cases}.$$

This correspondence has the property

$$\|a\| + \|b\| + \beta \cdot |\{t : a_t \geq \gamma\}| = \|\bar{a}\| + \|\bar{b}\| + \beta \cdot |\{t : \bar{b}_t \geq \gamma\}|. \tag{5.2.24c}$$

Third step. Each array $(\bar{a} \mid \bar{b})$ of the type (5.2.24b) uniquely corresponds to a path $\bar{P}$ from $(A_2 + 1, A_1 - 1)$ to (E_1, E_2). Besides there holds

$$\mathrm{ymaj}_{\beta;\gamma-1}(\bar{P}) = \|\bar{a}\| + \|\bar{b}\| + \beta \cdot |\{t : \bar{b}_t > \gamma - 1\}| - k(A_1 + A_2). \tag{5.2.25}$$

Now we are able to define $\Phi^{(2)}$. For any path P from $\mathcal{A}$ to $\mathcal{E}$ which crosses $x = y$ perform the steps 1–3. The image of P under application of $\Phi^{(2)}$ by definition is the resulting path $\bar{P}$,

$$\Phi^{(2)}(P) = \bar{P}.$$

Combining (5.2.23c), (5.2.24c), and (5.2.25), one obtains the desired property (5.2.21). Evidently, (5.2.22) results from (5.2.21) by setting $\beta = 0$.

The fact that the last $k - I$ elements in the rows of the arrays in (5.2.24a) are identical with the corresponding last $k - I$ elements in the array in (5.2.24b), implies that the last meeting point of P with $x = y + 1$ as well as the last meeting point of $\Phi^{(2)}(P)$ with $x = y + 1$ is $(b_I + 1, b_I)$, and that the portions of P and $\Phi^{(2)}(P)$ after this last meeting point are identical.

Finally we give an example to illustrate the action of $\Phi^{(2)}$. Consider the path $P_{\text{ex}} = ((-3,-1), 1121122111121222122211)$ in Figure 5, having the starting point $\mathcal{A}_{\text{ex}} = (-3,-1)$ and the final point $\mathcal{E}_{\text{ex}} = (9,9)$. (We have marked the North-East corners of the two paths in Figure 5 by thicker dots. So the paths should be easily distinguished.) In the first step P_{ex} is mapped to its array representation $(a^{(\text{ex})} \mid b^{(\text{ex})})$ given by

$$
\begin{array}{rll}
-3 \leq & -1\,1\,4\,5\,6\,7 & \leq 8 \\
0 \leq & 0\,2\,3\,5\,7\,9 & \leq 9
\end{array} .
$$

The index I which is maximal with respect to $a_{I+1}^{(\text{ex})} > b_I^{(\text{ex})}$ is $I = 4$. So in the second step, by (5.2.24) the array $(a^{(\text{ex})} \mid b^{(\text{ex})})$ is mapped to the array

$$
\begin{array}{rll}
0 \leq & 0\,2\,3\,5\,6\,7 & \leq 8 \\
-3 \leq & -1\,1\,4\,5\,7\,9 & \leq 9
\end{array} ,
$$

which is the array representation of the second path in Figure 5. S_{ex} is the last meeting point of P_{ex} and $\Phi^{(2)}(P_{\text{ex}})$ with the line $x = y + 1$.

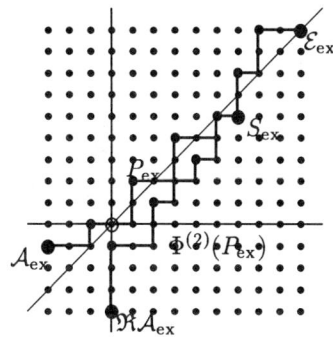

Figure 5

It is left to the reader to check that (5.2.21) always holds. □

Also for this case, a simpler bijection which satisfies (5.2.22) only has been previously given in [27, Proof of (5.37)]. It is also not extendable in order to be applicable with the strange major index.

5.3. A correspondence for pairs of intersecting lattice paths. Let $\mathcal{A} = (A_1, A_2)$, $\mathcal{B} = (B_1, B_2)$, $\mathcal{E} = (E_1, E_2)$, and $\mathcal{F} = (F_1, F_2)$ be lattice points in $\mathbb{Z}^2$ such that $A_1 \leq B_1$, $A_2 \geq B_2$, $E_1 \leq F_1$, $E_2 \geq F_2$. Recall [16, 17, 41] (see subsection 2.2) that the key for obtaining determinantal expressions for generating functions for families of nonintersecting lattice paths where the weight depends on the position of the edges of the paths in the lattice $\mathbb{Z}^2$ lies in the observation that the generating function for pairs (P_1, P_2) of *intersecting* paths, $P_1 : \mathcal{A} \to \mathcal{E}$, $P_2 : \mathcal{B} \to \mathcal{F}$, is equal to the generating function of *arbitrary* pairs (Q_1, Q_2) of paths, $Q_1 : \mathcal{B} \to \mathcal{E}$, $Q_2 : \mathcal{A} \to \mathcal{F}$. (Recall that paths are called *intersecting* if they have points in common, and *nonintersecting* if not.) This has already been explained in subsection 2.2. But we are not interested in edge weights but in weighing families of paths by the major index or the strange major indices, respectively. So what we need is a correspondence with similar properties, which in addition should take care of the major index or the strange major indices. The familiar interchanging procedure described in subsection 2.2 obviously does not.

The first observation is that there cannot be any useful result unless (A_1, A_2) and (B_1, B_2) lie on a line parallel to the anti-diagonal $x + y = 0$. After we have proved the following Proposition we shall try to give something like an "explanation" of this phenomenon. Here we want to content ourselves with an example. Take $\mathcal{A}_0 = (0, 0)$, $\mathcal{E}_0 = (2, 3)$, and $\mathcal{B}_0 = (1, -2)$, $\mathcal{F}_0 = (3, 1)$. The maj-generating function for intersecting pairs of paths (P_1, P_2), $P_1 : \mathcal{A}_0 \to \mathcal{E}_0$, $P_2 : \mathcal{B}_0 \to \mathcal{F}_0$, is given by

$$\sum_{\substack{P_1 : \mathcal{A}_0 \to \mathcal{E}_0 \\ P_2 : \mathcal{B}_0 \to \mathcal{F}_0 \\ P_1, P_2 \text{ intersect}}} q^{\text{maj } P_1 + \text{maj } P_2} = q^2 + 2q^3 + 4q^4 + 4q^5 + 5q^6 + 4q^7 + 2q^8 + q^9 + q^{10}. \quad (5.3.1)$$

Suppose that there would be some bijection taking care of the major index, such that the above intersecting pairs (P_1, P_2) of paths would uniquely correspond to pairs (Q_1, Q_2) of paths, $Q_1 : \mathcal{B}_0 \to \mathcal{E}_0$, $Q_2 : \mathcal{A}_0 \to \mathcal{F}_0$. Then the generating function (5.3.1) had to equal

$$\sum_{\substack{Q_1 : \mathcal{B}_0 \to \mathcal{E}_0 \\ Q_2 : \mathcal{A}_0 \to \mathcal{F}_0}} q^{\text{stat}_1 Q_1 + \text{stat}_2 Q_2} = \left(\sum_{Q_1 : \mathcal{B}_0 \to \mathcal{E}_0} q^{\text{stat}_1 Q_1} \right) \left(\sum_{Q_2 : \mathcal{A}_0 \to \mathcal{F}_0} q^{\text{stat}_2 Q_2} \right),$$

where stat_1 and stat_2 denote some major-like statistics. This means that the polynomial on the right-hand side should factor into two polynomials, one being the sum of 6 monomials of the form q^i, the second being the sum of 4 monomials of the form q^i, $i \in \{0, 1, \dots, 10\}$. (There are 6 paths from $\mathcal{B}_0 = (1, -2)$ to $\mathcal{E}_0 = (2, 3)$, and 4 paths from $\mathcal{A}_0 = (0, 0)$ to $\mathcal{F}_0 = (3, 1)$.) But the polynomial in (5.3.1) does not factor at all!

Therefore we will restrict ourselves to $\mathcal{A}$ and $\mathcal{B}$ lying on a line parallel to $x + y = 0$. This means that in the sequel we will always assume that there are some integers A, B, D such that $\mathcal{A} = (A + D, -A)$ and $\mathcal{B} = (B + D, -B)$. For this choice of the initial points of the paths it is indeed possible to find a strange major analogue (and hence also a major analogue) for the interchanging procedure of subsection 2.2. This analogue also maps pairs (P_1, P_2) of *intersecting* paths, $P_1 : \mathcal{A} \to \mathcal{E}$, $P_2 : \mathcal{B} \to \mathcal{F}$, onto *arbitrary* pairs (Q_1, Q_2) of paths, $Q_1 : \mathcal{B} \to \mathcal{E}$, $Q_2 : \mathcal{A} \to \mathcal{F}$, and it also preserves the path's portions after the last meeting point.

Proposition 27. FUNDAMENTAL BIJECTION FOR PAIRS OF INTERSECTING PATHS. Let $\mathcal{A} = (A + D, -A)$, $\mathcal{B} = (B + D, -B)$, $\mathcal{E} = (E_1, E_2)$, and $\mathcal{F} = (F_1, F_2)$ be lattice points in $\mathbb{Z}^2$ with $A < B$, $E_1 < F_1$, $E_2 \geq F_2$. There is a bijection Ψ between pairs (P_1, P_2), $P_1 : \mathcal{A} \to \mathcal{E}$, $P_2 : \mathcal{B} \to \mathcal{F}$, of intersecting paths and pairs (Q_1, Q_2), $Q_1 : \mathcal{B} \to \mathcal{E}$, $Q_2 : \mathcal{A} \to \mathcal{F}$, of paths such that the portions of P_1 and Q_1 after the last meeting point of P_1 and P_2 are identical, including this last meeting point, and *such that the portions of P_2 and Q_2 after this last meeting point, including the last meeting point, are identical. Moreover, if $(Q_1, Q_2) = \Psi\big((P_1, P_2)\big)$ then the equations*

$$\mathrm{ymaj}_{\beta;\gamma+1}(P_1) + \mathrm{ymaj}_{\beta;\gamma}(P_2) = \mathrm{ymaj}_{\beta;\gamma+1}(Q_1) + \mathrm{ymaj}_{\beta;\gamma}(Q_2) + B - A \quad (5.3.2)$$

and

$$\mathrm{xmaj}_{\beta;\gamma}(P_1) + \mathrm{xmaj}_{\beta;\gamma+1}(P_2) = \mathrm{xmaj}_{\beta;\gamma}(Q_1) + \mathrm{xmaj}_{\beta;\gamma+1}(Q_2) + B - A, \quad (5.3.3)$$

hold for all integers γ. In particular, there holds

$$\mathrm{maj}\, P_1 + \mathrm{maj}\, P_2 = \mathrm{maj}\, Q_1 + \mathrm{maj}\, Q_2 + B - A. \quad (5.3.4)$$

Besides, if $A + D < \gamma_1 \le B + D$ and $-B - 1 \le \gamma_2 < -A$, the equation

$$\mathrm{xmaj}_{\beta;\gamma_1}(P_1) + \mathrm{ymaj}_{\beta;\gamma_2}(P_2) = \mathrm{ymaj}_{\beta;\gamma_2+1}(Q_1) + \mathrm{xmaj}_{\beta;\gamma_1+1}(Q_2) + B - A + \beta \quad (5.3.5)$$

holds. Identity (5.3.5) also holds if the inequality $\gamma_1 \le B + D$ is replaced by the requirement that the last meeting point of P_1 and P_2 is located to the right or on the vertical line $x = \gamma_1$.

PROOF. Once the equation (5.3.2) is settled, the equations (5.3.3) and (5.3.4) immediately follow. For (5.3.3) this is seen using the relation (4.0.5) between ymaj and xmaj, while for deriving (5.3.4) we only have to set $\beta = 0$ in (5.3.2).

The construction of Ψ consists of four steps. As in the proof of Proposition 24, for the reader who is not interested in all the details we outline a short-cut through the following construction. The map Ψ is essentially explained by the lines (5.3.6a–e), (5.3.10), (5.3.11), (5.3.20), (5.3.21), and the first passage of the fourth step of the proof. For illustration, an example is carried out at the end of the proof.

First step. Again we utilize the array representation of paths. Let (P_1, P_2) be a pair of intersecting paths, $P_1 : \mathcal{A} \to \mathcal{E}$, $P_2 : \mathcal{B} \to \mathcal{F}$. P_1 and P_2 are in one-to-one correspondence with two-rowed arrays $(a \mid b)$ and $(c \mid d)$ of the types

$$\begin{array}{ccccc} A + D \le & a_1 & \dots & a_k & \le E_1 - 1 \\ -A + 1 \le & b_1 & \dots & b_k & \le E_2 \end{array} \quad (5.3.6a)$$

and

$$\begin{array}{ccccc} B + D \le & c_1 & \dots & c_l & \le F_1 - 1 \\ -B + 1 \le & d_1 & \dots & d_l & \le F_2 \end{array}, \quad (5.3.6b)$$

respectively.

Let $\mathcal{S}$ be the last meeting point of P_1 and P_2. By definition set $a_{k+1} := E_1$, $b_0 := -A$, $b_{k+1} := E_2 + 1$, and $c_{l+1} := F_1$. (Note that the thereby augmented sequences a, b, c remain strictly increasing.)

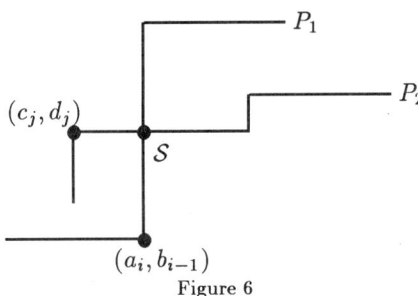

Figure 6

Considering the East-North corner (a_i, b_{i-1}) in P_1 immediately preceding $\mathcal{S}$ (and being allowed to be equal to $\mathcal{S}$) and the North-East corner (c_j, d_j) in P_2 immediately preceding $\mathcal{S}$ (and being allowed to be equal to $\mathcal{S}$), we get the inequalities (cf. Figure 6)

$$c_j \leq a_i < c_{j+1}, \tag{5.3.6c}$$

$$b_{i-1} \leq d_j < b_i, \tag{5.3.6d}$$

where

$$1 \leq i \leq k+1, \quad 1 \leq j \leq l. \tag{5.3.6e}$$

Of course, k, l, a_i, b_i, c_j, d_j, etc., refer to the array representations (5.3.6a,b) of P_1 and P_2. It now becomes apparent that the above assignments for $a_{k+1}, b_0, b_{k+1}, c_{l+1}$ are needed for the inequalities (5.3.6c,d) to make sense for $i = 1$, $i = k+1$, or $j = l$. Vice versa, if (5.3.6c,d,e) is satisfied then (a_i, d_j) is a meeting point between P_1 and P_2. Besides, a meeting point $\mathcal{S}$ is the *last* meeting point between P_1 and P_2 if and only if i *and* j are maximal with respect to the property (5.3.6c,d,e).

Summarizing, the existence of i, j satisfying (5.3.6c,d,e) characterize the array representations of *intersecting* pairs of paths. Thus we have a bijection between intersecting pairs (P_1, P_2) of paths, $P_1 : \mathcal{A} \to \mathcal{E}$, $P_2 : \mathcal{B} \to \mathcal{F}$, and pairs $((a \mid b), (c \mid d))$ of arrays of the types (5.3.6a) and (5.3.6b), respectively, for which there exist integers i, j which satisfy (5.3.6c,d,e). In the sequel we call such a pair of arrays a *pair of arrays of the type* (5.3.6).

We finally want to show (5.3.2). So it is necessary to relate $\mathrm{ymaj}_{\beta;\gamma+1}(P_1) + \mathrm{ymaj}_{\beta;\gamma}(P_2)$ to weights of the corresponding array representations (5.3.6a,b). Because of (4.0.6), according to the choice of γ, we have to distinguish between three cases:

Case A. $\gamma \geq -A - 1$. Here we have

$$\mathrm{ymaj}_{\beta;\gamma+1}(P_1) + \mathrm{ymaj}_{\beta;\gamma}(P_2) = \|a\| + \|b\| + \beta \cdot |\{t : b_t > \gamma + 1\}|$$
$$+ \|c\| + \|d\| + \beta \cdot |\{t : d_t > \gamma\}| - (k+l)D. \tag{5.3.7a}$$

Case B. $-A - 1 > \gamma \geq -B$. Here we have

$$\mathrm{ymaj}_{\beta;\gamma+1}(P_1) + \mathrm{ymaj}_{\beta;\gamma}(P_2) = \|a\| + \|b\| + \beta \cdot |\{t : a_t \geq D - \gamma - 1\}|$$
$$+ \|c\| + \|d\| + \beta \cdot |\{t : d_t > \gamma\}| - (k+l)D + (-A - \gamma - 1)\beta. \tag{5.3.7b}$$

Case C. $-B > \gamma$. Here we have

$$\mathrm{ymaj}_{\beta;\gamma+1}(P_1) + \mathrm{ymaj}_{\beta;\gamma}(P_2) = \|a\| + \|b\| + \beta \cdot |\{t : a_t \geq D - \gamma - 1\}|$$
$$+ \|c\| + \|d\| + \beta \cdot |\{t : c_t \geq D - \gamma\}| - (k+l)D + (-A - \gamma - 1)\beta + (-B - \gamma)\beta. \tag{5.3.7c}$$

To show (5.3.5) it will be necessary to use the following identity

$$\mathrm{xmaj}_{\beta;\gamma_1}(P_1) + \mathrm{ymaj}_{\beta;\gamma_2}(P_2) = \|a\| + \|b\| + \beta \cdot |\{t : a_t \geq \gamma_1\}|$$
$$+ \|c\| + \|d\| + \beta \cdot |\{t : d_t > \gamma_2\}| - (k+l)D, \tag{5.3.8a}$$

which in view of (4.0.6) and (4.0.7) certainly holds if $A + D < \gamma_1$ and $-B \leq \gamma_2$. For $\gamma_2 = -B - 1$ we have

$$\mathrm{xmaj}_{\beta;\gamma_1}(P_1) + \mathrm{ymaj}_{\beta;-B-1}(P_2) = \|a\| + \|b\| + \beta \cdot |\{t : a_t \geq \gamma_1\}|$$
$$+ \|c\| + \|d\| + \beta \cdot |\{t : c_t \geq B + D + 1\}| - (k+l)D + \beta. \tag{5.3.8b}$$

Second step. This is the crucial step in the construction of Ψ.

Let I and J be maximal with

$$c_J \leq a_I < c_{J+1}, \tag{5.3.9a}$$
$$b_{I-1} \leq d_J < b_I, \tag{5.3.9b}$$

and

$$1 \leq I \leq k+1, \quad 1 \leq J \leq l. \tag{5.3.9c}$$

We map a pair $((a \mid b), (c \mid d))$ of arrays of the type (5.3.6) to the pair $((\bar{a} \mid \bar{b}), (\bar{c} \mid \bar{d}))$ of arrays, where $(\bar{a} \mid \bar{b}) = (\bar{a}_1, \ldots, \bar{a}_{k-I+J+1} \mid \bar{b}_2, \ldots, \bar{b}_{k-I+J+1})$ is given by

$$\begin{array}{cccccc} c_1 - 1 & c_2 - 1 & \ldots & c_J - 1 & a_I & \ldots & a_k \\ d_1 + 1 & \ldots & d_{J-1} + 1 & b_I & \ldots & b_k \end{array} \tag{5.3.10a}$$

and $(\bar{c} \mid \bar{d}) = (\bar{c}_2, \ldots, \bar{c}_{l-J+I} \mid \bar{d}_1, \ldots, \bar{d}_{l-J+I})$ is given by

$$\begin{array}{cccccc} a_1 + 1 & \ldots \ldots \ldots & a_{I-1} + 1 & c_{J+1} & \ldots & c_l \\ b_1 - 1 & b_2 - 1 & \ldots & b_{I-1} - 1 & d_J & d_{J+1} & \ldots & d_l \end{array}. \tag{5.3.10b}$$

We claim that this mapping is invertible. More precisely, we are going to establish that the mapping (5.3.10) is a bijection between pairs $((a \mid b), (c \mid d))$ of arrays of the type (5.3.6) and pairs $((\bar{a} \mid \bar{b}), (\bar{c} \mid \bar{d}))$ of arrays of the types

$$\begin{array}{ccccc} B + D - 1 \leq & \bar{a}_1 & \ldots \ldots & \bar{a}_{\bar{k}} & \leq E_1 - 1 \\ -B + 2 \leq & & \bar{b}_2 & \ldots & \bar{b}_{\bar{k}} & \leq E_2 \end{array} \tag{5.3.11a}$$

and

$$\begin{array}{ccccc} A + D + 1 \leq & & \bar{c}_2 & \ldots & \bar{c}_{\bar{l}} & \leq F_1 - 1 \\ -A \leq & \bar{d}_1 & \ldots \ldots & \bar{d}_{\bar{l}} & \leq F_2 \end{array}, \tag{5.3.11b}$$

respectively. That the bounds in (5.3.11) are the right ones, is trivial.

To establish that (5.3.10) is invertible we construct the inverse mapping. Suppose that $((\bar{a} \mid \bar{b}), (\bar{c} \mid \bar{d}))$ is a pair of arrays of the type (5.3.11). By definition set $\bar{a}_{\bar{k}+1} := E_1$, $\bar{b}_1 := -B+1$, $\bar{b}_{\bar{k}+1} := E_2+1$, $\bar{c}_1 := A+D$, and $\bar{c}_{\bar{l}+1} := F_1$. (Again note that these assignments guarantee the thereby augmented sequences $\bar{a}, \bar{b}, \bar{c}$ to remain strictly increasing.)

Let $\bar{I}$ and $\bar{J}$ be maximal with

$$\bar{c}_{\bar{J}} \leq \bar{a}_{\bar{I}} < \bar{c}_{\bar{J}+1}, \tag{5.3.12a}$$

$$\bar{b}_{\bar{I}-1} \leq \bar{d}_{\bar{J}} < \bar{b}_{\bar{I}}, \tag{5.3.12b}$$

and

$$2 \leq \bar{I} \leq \bar{k}+1, \quad 1 \leq \bar{J} \leq \bar{l}. \tag{5.3.12c}$$

The assignments for $\bar{a}_{\bar{k}+1}, \bar{b}_1$, etc., have been necessary in order that the inequalities (5.3.12a,b) make sense for $\bar{I} = 2$, $\bar{I} = \bar{k}+1$, $\bar{J} = 1$, or $\bar{J} = \bar{l}$. We map a pair $((\bar{a} \mid \bar{b}), (\bar{c} \mid \bar{d}))$ of arrays of the type (5.3.11) to the arrays

$$\begin{matrix} \bar{c}_2 - 1 & \cdots & \bar{c}_{\bar{J}} - 1 & \bar{a}_{\bar{I}} & \cdots & \bar{a}_{\bar{k}} \\ \bar{d}_1 + 1 & \cdots & \bar{d}_{\bar{J}-1} & \bar{b}_{\bar{I}} & \cdots & \bar{b}_{\bar{k}} \end{matrix} \tag{5.3.13a}$$

and

$$\begin{matrix} \bar{a}_1 + 1 & \cdots\cdots\cdots & \bar{a}_{\bar{I}-1} + 1 & \bar{c}_{\bar{J}+1} & \cdots & \bar{c}_{\bar{l}} \\ \bar{b}_2 - 1 & \cdots & \bar{b}_{\bar{I}-1} - 1 & \bar{d}_{\bar{J}} & \bar{d}_{\bar{J}+1} & \cdots & \bar{d}_{\bar{l}} \end{matrix}. \tag{5.3.13b}$$

The mapping (5.3.13) will be shown to be the inverse of (5.3.10).

The first thing to have to be established, is that the map (5.3.13) is always well-defined. This is the case if and only if there exist $\bar{I}$ and $\bar{J}$ which are maximal with respect to (5.3.12). So we have to show that there is at least one pair of integers i, j, with

$$\bar{c}_j \leq \bar{a}_i < \bar{c}_{j+1}, \tag{5.3.14a}$$

$$\bar{b}_{i-1} \leq \bar{d}_j < \bar{b}_i, \tag{5.3.14b}$$

and

$$2 \leq i \leq \bar{k}+1, \quad 1 \leq j \leq \bar{l}. \tag{5.3.14c}$$

Let $U = \{2, \ldots, \bar{k}+1\}$ be the range of the index i and $V = \{1, \ldots, \bar{l}\}$ be the range of the index j. Next, two mappings f and g, $f : U \to V$, $g : V \to U$, are defined. For $i \in U$ denote by $f(i)$ the uniquely determined integer that satisfies

$$\bar{c}_{f(i)} \leq \bar{a}_i < \bar{c}_{f(i)+1}. \tag{5.3.15}$$

Likewise, for $j \in V$ denote by $g(j)$ the uniquely determined integer that satisfies

$$\bar{b}_{g(j)-1} \leq \bar{d}_j < \bar{b}_{g(j)}. \tag{5.3.16}$$

We have to confirm that f and g are well-defined for all $i \in U$, and that g is well-defined for all $j \in V$. What has to be established is the existence of integers $f(i)$ satisfying (5.3.15) and the existence of integers $g(j)$ satisfying (5.3.16). Once this has been done the uniqueness is clear from the fact that the sequences $\bar{a}, \bar{b}, \bar{c}, \bar{d}$ are in strictly increasing order.

So let $i \in U$. We have $B + D - 1 \leq \bar{a}_1 < \bar{a}_i \leq \bar{a}_{\bar{k}+1} = E_1$. Because of the assumptions $A < B$ and $E_1 < F_1$, we conclude that $A + D \leq \bar{a}_i < F_1$, or in terms of $\bar{c}_1$ and $\bar{c}_{\bar{l}+1}$, $\bar{c}_1 \leq \bar{a}_i < \bar{c}_{\bar{l}+1}$. Clearly this settles the existence of an integer $f(i)$ satisfying (5.3.15).

Similarly, let $j \in V$. We have $-A \leq \bar{d}_j \leq F_2$. Because of the assumptions $A < B$ and $E_2 \geq F_2$, we get $-B + 1 \leq \bar{d}_j < E_2 + 1$. This inequality, written in terms of $\bar{b}_1$ and $\bar{b}_{\bar{k}+1}$, reads $\bar{b}_1 \leq \bar{d}_j < \bar{b}_{\bar{k}+1}$, which confirms the existence of an integer $g(j)$ satisfying (5.3.16).

It is a simple matter of fact that f and g are monotone functions. To be precise, if $i_1 \leq i_2$ then $f(i_1) \leq f(i_2)$, and if $j_1 \leq j_2$ then $g(j_1) \leq g(j_2)$. This is due to the strictly increasing order of $\bar{a}, \bar{b}, \bar{c}$, and $\bar{d}$. Now start with $2 \in U$ and alternatively apply f and g. To be precise, we form the sequence $s_1, s_2, s_3, \ldots$, where $s_1 = 2$, $s_{2t} = f(s_{2t-1})$, and $s_{2t+1} = g(s_{2t})$, $t = 1, 2, \ldots$. Then we consider the subsequences $s_1, s_3, s_5, \ldots$ and $s_2, s_4, s_6, \ldots$. Obviously, because of the monotonicity of both f and g these subsequences are weakly increasing. In addition they are infinite and bounded sequences, hence both must eventually become stationary. Denote by $\bar{i}$ the limit element of the first sequence, and by $\bar{j}$ the limit element of the second sequence. By definition of the sequences we have $f(\bar{i}) = \bar{j}$ and $g(\bar{j}) = \bar{i}$ which by means of (5.3.15) and (5.3.16) says that $\bar{c}_{\bar{j}} \leq \bar{a}_{\bar{i}} < \bar{c}_{\bar{j}+1}$ and $\bar{b}_{\bar{i}-1} \leq \bar{d}_{\bar{j}} < \bar{b}_{\bar{i}}$. Hence we have found a pair of indices which satisfies (5.3.14).

Now that we have convinced ourselves that the mapping (5.3.13) is well-defined, it remains to show that it is indeed the inverse of (5.3.10). As before, let $((a \mid b)$, $(c \mid d))$ be a pair of arrays of the type (5.3.6) and I, J be maximal with respect to (5.3.9). Denote the array in (5.3.10a) by $(\bar{a} \mid \bar{b})$ and the array in (5.3.10b) by $(\bar{c} \mid \bar{d})$. (Note that $\bar{k} = k - I + J + 1$ and $\bar{l} = l - J + I$.) Let $\bar{I}, \bar{J}$ be maximal with respect to (5.3.12). All we have to show is that $\bar{I} = J + 1$ and $\bar{J} = I$ since then from (5.3.10) it is obvious that by applying (5.3.13) to the pair $((\bar{a} \mid \bar{b}), (\bar{c} \mid \bar{d}))$ we obtain the pair $((a \mid b), (c \mid d))$ again.

Also from (5.3.9) and (5.3.10) it is clear that (5.3.12) holds if $\bar{I}$ is replaced by $J + 1$ and $\bar{J}$ by I. Hence the actual values of $\bar{I}, \bar{J}$ because of their maximality must be greater or equal. To be precisely, the inequalities $\bar{I} \geq J + 1$ and $\bar{J} \geq I$ must hold. We distinguish between three cases.

Case 1. $\bar{I} > J + 1$ and $\bar{J} > I$. Under these assumptions by (5.3.10) we have $\bar{a}_{\bar{I}} = a_{\bar{I}-J+I-1}$, $\bar{b}_{\bar{I}} = b_{\bar{I}-J+I-1}$, $\bar{b}_{\bar{I}-1} = b_{\bar{I}-J+I-2}$, $\bar{c}_{\bar{J}} = c_{\bar{J}-I+J}$, $\bar{c}_{\bar{J}+1} = c_{\bar{J}-I+J+1}$, and $\bar{d}_{\bar{J}} = d_{\bar{J}-I+J}$. Setting $I_0 := \bar{I} - J + I - 1$ and $J_0 := \bar{J} - I + J$, the inequalities (5.3.12a,b) read $c_{J_0} \leq a_{I_0} < c_{J_0+1}$ and $b_{I_0-1} \leq d_{J_0} < b_{I_0}$, respectively. Besides, because of $J + 1 < \bar{I} \leq \bar{k} + 1 = k - I + J + 2$ we get $I < I_0 \leq k + 1$, and because of $I < \bar{J} \leq \bar{l} = l - J + I$ we get $J < J_0 \leq l$. This contradicts the maximality of I and J with respect to (5.3.9).

Case 2. $\bar{I} > J + 1$ and $\bar{J} = I$. Here by (5.3.10) we have $\bar{c}_{\bar{J}+1} = c_{J+1}$, $\bar{d}_{\bar{J}} = d_J$, and $\bar{a}_{\bar{I}}, \bar{b}_{\bar{I}}, \bar{b}_{\bar{I}-1}$ as in Case 1. Setting $I_0 := \bar{I} - J + I - 1$ and $J_0 := J$ from the second inequality in (5.3.12a) we get $a_{I_0} < c_{J_0+1}$, while (5.3.12b) reads $b_{I_0-1} \leq d_{J_0} < b_{I_0}$. Again, because of $J + 1 < \bar{I} \leq \bar{k} + 1 = k - I + J + 2$ we have $I < I_0 \leq k + 1$. In particular, from this and (5.3.9a) we conclude that $c_J \leq a_I < a_{I_0}$. This implies $c_{J_0} < a_{I_0} < c_{J_0+1}$. This is a contradiction to the maximality of I and J with respect to (5.3.9).

Case 3. $\bar{I} = J + 1$ and $\bar{J} > I$. Here by (5.3.10) we have $\bar{a}_{\bar{I}} = a_I$, $\bar{b}_{\bar{I}} = b_I$, and $\bar{c}_{\bar{J}}, \bar{c}_{\bar{J}+1}, \bar{d}_{\bar{J}}$ as in Case 1. Setting $I_0 := I$ and $J_0 := \bar{J} - I + J$, the inequality (5.3.12a) reads $c_{J_0} \leq a_{I_0} < c_{J_0+1}$, while from the second inequality in (5.3.12b) we derive $d_{J_0} < b_{I_0}$. Because of $I < \bar{J} \leq \bar{l} = l - J + I$ we have $J < J_0 \leq l$. In particular, from this and (5.3.9b) we deduce $b_{I-1} \leq d_J < d_{J_0}$. This implies $b_{I_0-1} < d_{J_0} < b_{I_0}$. Again we obtain a contradiction to the maximality of I and J with respect to (5.3.9).

This completes the proof that (5.3.10) is a bijection between pairs of arrays of the type (5.3.6) and pairs of arrays of the type (5.3.11).

Finally we need to say something about weight properties. Again, let $((a \mid b), (c \mid d))$ be a pair of arrays of the type (5.3.6) and denote by $((\bar{a} \mid \bar{b}), (\bar{c} \mid \bar{d}))$ its image under application of (5.3.10). The right-hand sides of (5.3.7a,b,c) serve as a guide which expressions have to be considered. To begin with, a short computation shows that

$$\|a\| + \|b\| + \|c\| + \|d\| - (k+l)D = \|\bar{a}\| + \|\bar{b}\| + \|\bar{c}\| + \|\bar{d}\| - (\bar{k} + \bar{l})D + D + 1. \quad (5.3.17)$$

To count contributed β's, even five cases have to be distinguished. Since the sense of the following manipulations will only become clear later, maybe for the moment the reader should skip the following case study and come back to it before reading the passage after (5.3.21), when the study of weight properties is continued.

Case A1. $\gamma \geq -A$. Having in mind (5.3.7a) we obtain

$$\begin{aligned}
\beta \cdot &|\{t : b_t > \gamma + 1\}| + \beta \cdot |\{t : d_t > \gamma\}| \\
&= \beta \cdot |\{t : b_t > \gamma + 1 \text{ and } t \geq I\}| + \beta \cdot |\{t : b_t - 1 > \gamma \text{ and } t < I\}| \\
&\quad + \beta \cdot |\{t : d_t > \gamma \text{ and } t \geq J\}| + \beta \cdot |\{t : d_t + 1 > \gamma + 1 \text{ and } t < J\}| \\
&= \beta \cdot |\{t : \bar{b}_t > \gamma + 1\}| + \beta \cdot |\{t : \bar{d}_t > \gamma\}|. \quad (5.3.18a)
\end{aligned}$$

Case A2. $\gamma = -A - 1$. Again having in mind (5.3.7a) we get

$$\begin{aligned}
\beta \cdot &|\{t : b_t > -A\}| + \beta \cdot |\{t : d_t > -A - 1\}| \\
&= \beta \cdot |\{t : b_t > -A \text{ and } t \geq I\}| + \beta \cdot |\{t : b_t > -A \text{ and } t < I\}| \\
&\quad + \beta \cdot |\{t : d_t > -A - 1 \text{ and } t \geq J\}| + \beta \cdot |\{t : d_t > -A - 1 \text{ and } t < J\}|.
\end{aligned}$$

Since because of (5.3.6a) the relations $b_t > -A$ and $a_t \geq A + D$ hold for all $t = 1, \ldots, k$, in the above equation the condition $b_t > -A$ may freely be substituted by $a_t + 1 \geq$

$A + D + 1$. Another observation is that by (5.3.9b) we have $-A = b_0 \leq b_{I-1} \leq d_J$. On the other hand (5.3.9a) gives $A + D \leq a_I < c_{J+1}$. These facts imply

$$|\{t : d_t > -A - 1 \text{ and } t \geq J\}| = l - J + 1$$
$$= |\{t : c_t \geq A + D + 1 \text{ and } t \geq J + 1\}| + 1.$$

Putting everything together, we obtain

$$\beta \cdot |\{t : b_t > -A\}| + \beta \cdot |\{t : d_t > -A - 1\}|$$
$$= \beta \cdot |\{t : b_t > -A \text{ and } t \geq I\}| + \beta \cdot |\{t : a_t + 1 \geq A + D + 1 \text{ and } t < I\}|$$
$$+ \beta \cdot |\{t : c_t \geq A + D + 1 \text{ and } t \geq J + 1\}| + \beta$$
$$+ \beta \cdot |\{t : d_t + 1 > -A \text{ and } t < J\}|$$
$$= \beta \cdot |\{t : \bar{b}_t > -A\}| + \beta \cdot |\{t : \bar{c}_t \geq A + D + 1\}| + \beta. \tag{5.3.18b}$$

Case B. $-A - 1 > \gamma \geq -B$. This time having in mind (5.3.7b), we have

$$\beta \cdot |\{t : a_t \geq D - \gamma - 1\}| + \beta \cdot |\{t : d_t > \gamma\}|$$
$$= \beta \cdot |\{t : a_t \geq D - \gamma - 1 \text{ and } t \geq I\}| + \beta \cdot |\{t : a_t \geq D - \gamma - 1 \text{ and } t < I\}|$$
$$+ \beta \cdot |\{t : d_t > \gamma \text{ and } t \geq J\}| + \beta \cdot |\{t : d_t > \gamma \text{ and } t < J\}|.$$

Use of $\gamma \geq -B$ and of (5.3.9a) gives $D - \gamma \leq B + D \leq c_J \leq a_I$. On the other hand obviously we have $\gamma + 1 < -A \leq b_I$. These facts imply

$$|\{t : a_t \geq D - \gamma - 1 \text{ and } t \geq I\}| = k - I + 1$$
$$= |\{t : b_t > \gamma + 1 \text{ and } t \geq I\}|.$$

Moreover, by $-A - 1 > \gamma$ and (5.3.9b) we get $\gamma < -A = b_0 \leq b_{I-1} \leq d_J$, and by $\gamma \geq -B$ we get $D - \gamma \leq B + D \leq c_J$. These facts imply

$$|\{t : d_t > \gamma \text{ and } t \geq J\}| = l - J + 1$$
$$= |\{t : c_t \geq D - \gamma \text{ and } t \geq J + 1\}| + 1.$$

Putting everything together, we obtain

$$\beta \cdot |\{t : a_t \geq D - \gamma - 1\}| + \beta \cdot |\{t : d_t > \gamma\}|$$
$$= \beta \cdot |\{t : b_t > \gamma + 1 \text{ and } t \geq I\}| + \beta \cdot |\{t : a_t + 1 \geq D - \gamma \text{ and } t < I\}|$$
$$+ \beta \cdot |\{t : c_t \geq D - \gamma \text{ and } t \geq J + 1\}| + \beta$$
$$+ \beta \cdot |\{t : d_t + 1 > \gamma + 1 \text{ and } t < J\}|$$
$$= \beta \cdot |\{t : \bar{b}_t > \gamma + 1\}| + \beta \cdot |\{t : \bar{c}_t \geq D - \gamma\}| + \beta. \tag{5.3.18c}$$

Case C1. $\gamma = -B - 1$. Having in mind (5.3.7c), we get

$$\beta \cdot |\{t : a_t \geq B + D\}| + \beta \cdot |\{t : c_t \geq B + D + 1\}|$$
$$= \beta \cdot |\{t : a_t \geq B + D \text{ and } t \geq I\}| + \beta \cdot |\{t : a_t \geq B + D \text{ and } t < I\}|$$
$$+ \beta \cdot |\{t : c_t \geq B + D + 1 \text{ and } t \geq J + 1\}|$$
$$+ \beta \cdot |\{t : c_t \geq B + D + 1 \text{ and } t < J + 1\}|.$$

By (5.3.9a) we have $B + D \leq c_J \leq a_I$, while by (5.3.9b) we have $-B < d_J < b_I$. This yields

$$|\{t : a_t \geq B + D \text{ and } t \geq I\}| = k - I + 1$$
$$= |\{t : b_t > -B \text{ and } t \geq I\}|.$$

Moreover, by definition for all t there hold $c_t \geq B + D$ and $d_t > -B$. Hence

$$|\{t : c_t \geq B + D + 1 \text{ and } t < J + 1\}| = J - \chi(c_1 = B + D)$$
$$= |\{t : d_t + 1 > -B \text{ and } t < J\}| + 1 - \chi(c_1 = B + D),$$

where, as in section 2, χ is the usual truth function. Putting everything together, we obtain

$$\beta \cdot |\{t : a_t \geq B + D\}| + \beta \cdot |\{t : c_t \geq B + D + 1\}|$$
$$= \beta \cdot |\{t : b_t > -B \text{ and } t \geq I\}| + \beta \cdot |\{t : a_t + 1 \geq B + D + 1 \text{ and } t < I\}|$$
$$+ \beta \cdot |\{t : c_t \geq B + D + 1 \text{ and } t \geq J + 1\}|$$
$$+ \beta \cdot |\{t : d_t + 1 > -B \text{ and } t < J\}| + (1 - \chi(c_1 = B + D)) \cdot \beta$$
$$= \beta \cdot |\{t : \bar{b}_t > -B\}| + \beta \cdot |\{t : \bar{c}_t \geq B + D + 1\}| + \chi(c_1 > B + D) \cdot \beta.$$
$$(5.3.18d)$$

Case C2. $\gamma < -B - 1$. Once more, having in mind (5.3.7c), we get

$$\beta \cdot |\{t : a_t \geq D - \gamma - 1\}| + \beta \cdot |\{t : c_t \geq D - \gamma\}|$$
$$= \beta \cdot |\{t : a_t \geq D - \gamma - 1 \text{ and } t \geq I\}| + \beta \cdot |\{t : a_t + 1 \geq D - \gamma \text{ and } t < I\}|$$
$$+ \beta \cdot |\{t : c_t \geq D - \gamma \text{ and } t \geq J + 1\}|$$
$$+ \beta \cdot |\{t : c_t - 1 \geq D - \gamma - 1 \text{ and } t < J + 1\}|$$
$$= \beta \cdot |\{t : \bar{a}_t \geq D - \gamma - 1\}| + \beta \cdot |\{t : \bar{c}_t \geq D - \gamma\}|.$$
$$(5.3.18e)$$

In order to establish (5.3.5), we need the following considerations. For $\gamma_2 \geq -B$, having in mind (5.3.8a), we have

$$\beta \cdot |\{t : a_t \geq \gamma_1\}| + \beta \cdot |\{t : d_t > \gamma_2\}|$$
$$= \beta \cdot |\{t : a_t \geq \gamma_1 \text{ and } t \geq I\}| + \beta \cdot |\{t : a_t \geq \gamma_1 \text{ and } t < I\}|$$
$$+ \beta \cdot |\{t : d_t > \gamma_2 \text{ and } t \geq J\}| + \beta \cdot |\{t : d_t > \gamma_2 \text{ and } t < J\}|.$$

If $\gamma_1 \leq B + D$ then using (5.3.9a) we have $\gamma_1 \leq B + D \leq c_J \leq a_I$. On the other hand, if the last meeting point between P_1 and P_2 is located to the right or on the line $x = \gamma_1$, we directly infer $\gamma_1 \leq a_I$. (Recall that this last meeting point actually is (a_I, d_J).) Using this, (5.3.9a,b), and the assumption $\gamma_2 < -A$, we get the inequalities $\gamma_1 \leq a_I < c_{J+1}$, $\gamma_2 + 1 \leq -A < b_I$, and $\gamma_2 < -A \leq b_{I-1} \leq d_J$. These facts imply

$$|\{t : a_t \geq \gamma_1 \text{ and } t \geq I\}| = k - I + 1$$
$$= |\{t : b_t > \gamma_2 + 1 \text{ and } t \geq I\}|$$

and

$$|\{t : d_t > \gamma_2 \text{ and } t \geq J\}| = l - J + 1$$
$$= |\{t : c_t \geq \gamma_1 + 1 \text{ and } t \geq J + 1\}| + 1.$$

Putting everything together, we obtain

$$\beta \cdot |\{t : a_t \geq \gamma_1\}| + \beta \cdot |\{t : d_t > \gamma_2\}|$$
$$= \beta \cdot |\{t : b_t > \gamma_2 + 1 \text{ and } t \geq I\}| + \beta \cdot |\{t : a_t + 1 \geq \gamma_1 + 1 \text{ and } t < I\}|$$
$$+ \beta \cdot |\{t : c_t \geq \gamma_1 + 1 \text{ and } t \geq J + 1\}| + \beta$$
$$+ \beta \cdot |\{t : d_t + 1 > \gamma_2 + 1 \text{ and } t < J\}|$$
$$= \beta \cdot |\{t : \bar{b}_t > \gamma_2 + 1\}| + \beta \cdot |\{t : \bar{c}_t \geq \gamma_1 + 1\}| + \beta. \tag{5.3.19a}$$

On the other hand, for $\gamma_2 = -B - 1$, we have to consider the right-hand side of (5.3.8b). We have

$$\beta \cdot |\{t : a_t \geq \gamma_1\}| + \beta \cdot |\{t : c_t \geq B + D + 1\}|$$
$$= \beta \cdot |\{t : a_t \geq \gamma_1 \text{ and } t \geq I\}| + \beta \cdot |\{t : a_t \geq \gamma_1 \text{ and } t < I\}|$$
$$+ \beta \cdot |\{t : c_t \geq B + D + 1 \text{ and } t \geq J + 1\}|$$
$$+ \beta \cdot |\{t : c_t \geq B + D + 1 \text{ and } t < J + 1\}|.$$

Examining the above considerations for the case $\gamma_2 \geq -B$, we discover that the inequality $\gamma_1 \geq -B$ in fact has never been used. So also here there hold $\gamma_1 \leq a_I < c_{J+1}$ and $-B < b_I$. Besides we have

$$|\{t : c_t \geq B + D + 1 \text{ and } t < J + 1\}| = J - \chi(c_1 = B + D)$$
$$= |\{t : d_t + 1 > -B \text{ and } t < J\}| + 1 - \chi(c_1 = B + D).$$

Putting everything together, we obtain

$$\beta \cdot |\{t : a_t \geq \gamma_1\}| + \beta \cdot |\{t : c_t \geq B + D + 1\}|$$
$$= \beta \cdot |\{t : b_t > -B \text{ and } t \geq I\}| + \beta \cdot |\{t : a_t + 1 \geq \gamma_1 + 1 \text{ and } t < I\}|$$
$$+ \beta \cdot |\{t : c_t \geq \gamma_1 + 1 \text{ and } t \geq J + 1\}|$$
$$+ \beta \cdot |\{t : d_t + 1 > -B \text{ and } t < J\}| + (1 - \chi(c_1 = B + D)) \cdot \beta$$
$$= \beta \cdot |\{t : \bar{b}_t > -B\}| + \beta \cdot |\{t : \bar{c}_t \geq \gamma_1 + 1\}| + \chi(c_1 > B + D) \cdot \beta. \tag{5.3.19b}$$

Third step. So far we have bijectively mapped intersecting pairs (P_1, P_2) of paths onto pairs $((\bar{a} \mid \bar{b}), (\bar{c} \mid \bar{d}))$ of arrays of the type (5.3.11). To interpret these arrays as paths again, we have to modify them such that both become arrays consisting of rows with equal length. For $(\bar{a} \mid \bar{b})$ of the type (5.3.11a) we distinguish between the cases $\bar{a}_1 = B + D - 1$ and $\bar{a}_1 > B + D - 1$. In the first case $\bar{a}_1$ is deleted from the array $(\bar{a} \mid \bar{b})$, in the second case the element $-B + 1$ is inserted as first element in the second row. This yields a new array $(\hat{a} \mid \hat{b})$ which comes out of $(\bar{a} \mid \bar{b})$ by

$$
(\hat{a} \mid \hat{b}) = \begin{cases} \begin{array}{cccc} \bar{a}_2 & \ldots & \bar{a}_{\bar{k}} \\ \bar{b}_2 & \ldots & \bar{b}_{\bar{k}} \end{array} & \bar{a}_1 = B + D - 1 \\ \begin{array}{cccc} \bar{a}_1 & \bar{a}_2 & \ldots & \bar{a}_{\bar{k}} \\ -B+1 & \bar{b}_2 & \ldots & \bar{b}_{\bar{k}} \end{array} & \bar{a}_1 > B + D - 1 \, . \end{cases} \tag{5.3.20a}
$$

Similarly, we define a new array $(\hat{c} \mid \hat{d})$ which comes out of $(\bar{c} \mid \bar{d})$ by

$$
(\hat{c} \mid \hat{d}) = \begin{cases} \begin{array}{cccc} \bar{c}_2 & \ldots & \bar{c}_{\bar{l}} \\ \bar{d}_2 & \ldots & \bar{d}_{\bar{l}} \end{array} & \bar{d}_1 = -A \\ \begin{array}{cccc} A+D & \bar{c}_2 & \ldots & \bar{c}_{\bar{l}} \\ \bar{d}_1 & \bar{d}_2 & \ldots & \bar{d}_{\bar{l}} \end{array} & \bar{d}_1 > -A \, . \end{cases} \tag{5.3.20b}
$$

It is clear that (5.3.20) defines a bijection between pairs $((\bar{a} \mid \bar{b}), (\bar{c} \mid \bar{d}))$ of the type (5.3.11) and pairs $((\hat{a} \mid \hat{b}), (\hat{c} \mid \hat{d}))$ of the types

$$
\begin{array}{ccccc} B + D \leq & \hat{a}_1 & \ldots & \hat{a}_{\hat{k}} & \leq E_1 - 1 \\ -B + 1 \leq & \hat{b}_1 & \ldots & \hat{b}_{\hat{k}} & \leq E_2 \end{array} \tag{5.3.21a}
$$

and

$$
\begin{array}{ccccc} A + D \leq & \hat{c}_1 & \ldots & \hat{c}_{\hat{l}} & \leq F_1 - 1 \\ -A + 1 \leq & \hat{d}_1 & \ldots & \hat{d}_{\hat{l}} & \leq F_2 \end{array} \,, \tag{5.3.21b}
$$

respectively.

Turning to the weights of the arrays, first observe that in all four cases of (5.3.20) the following identity holds,

$$
\|\bar{a}\| + \|\bar{b}\| + \|\bar{c}\| + \|\bar{d}\| - (\bar{k} + \bar{l})D
$$
$$
= \|\hat{a}\| + \|\hat{b}\| + \|\hat{c}\| + \|\hat{d}\| - (\hat{k} + \hat{l})D + B - A - D - 1. \tag{5.3.22}
$$

Besides, it is readily seen that (5.3.18a,b,c,e) also hold if $\bar{a}, \bar{b}, \bar{c}, \bar{d}$ are replaced by $\hat{a}, \hat{b}, \hat{c}, \hat{d}$, respectively. In the exceptional case C1 (i.e. $\gamma = -B - 1$) we have

$$
|\{t : \hat{b}_t > -B\}| = \bar{k} - 1 + \chi(\bar{a}_1 > B + D - 1)
$$
$$
= |\{t : \bar{b}_t > -B\}| + \chi(c_1 > B + D) \tag{5.3.23}
$$

by remembering (5.3.10a). Combining this with (5.3.18d) gives

$$
\beta \cdot |\{t : a_t \geq B + D\}| + \beta \cdot |\{t : c_t \geq B + D + 1\}|
$$
$$
= \beta \cdot |\{t : \hat{b}_t > -B\}| + \beta \cdot |\{t : \hat{c}_t \geq B + D + 1\}|. \tag{5.3.24}
$$

In order to finally prove (5.3.5), we observe that for $\gamma_2 \geq -B$ the identity (5.3.19a) also holds if $\bar{b}, \bar{c}$ are replaced by $\hat{b}, \hat{c}$, respectively. For $\gamma_2 = -B - 1$ however, we may use (5.3.23) to get from (5.3.19b)

$$\beta \cdot |\{t : a_t \geq \gamma_1\}| + \beta \cdot |\{t : c_t \geq B + D + 1\}|$$
$$\beta \cdot |\{t : \hat{b}_t > -B\}| + \beta \cdot |\{t : \hat{c}_t \geq \gamma_1 + 1\}|. \quad (5.3.25)$$

Fourth step. Obviously, pairs $((\hat{a} \mid \hat{b}), (\hat{c} \mid \hat{d}))$ of arrays of the type (5.3.21) uniquely correspond to pairs (Q_1, Q_2) of paths, $Q_1 : \mathcal{B} \to \mathcal{E}$, $Q_2 : \mathcal{A} \to \mathcal{F}$ such that the following weight properties hold:

Case A'. $\gamma \geq -A$. Here we have

$$\mathrm{ymaj}_{\beta;\gamma+1}(Q_1) + \mathrm{ymaj}_{\beta;\gamma}(Q_2) = \|\hat{a}\| + \|\hat{b}\| + \beta \cdot |\{t : \hat{b}_t > \gamma + 1\}|$$
$$+ \|\hat{c}\| + \|\hat{d}\| + \beta \cdot |\{t : \hat{d}_t > \gamma\}| - (\hat{k} + \hat{l})D. \quad (5.3.26a)$$

Case B'. $-A > \gamma \geq -B - 1$. Here we have

$$\mathrm{ymaj}_{\beta;\gamma+1}(Q_1) + \mathrm{ymaj}_{\beta;\gamma}(Q_2) = \|\hat{a}\| + \|\hat{b}\| + \beta \cdot |\{t : \hat{b}_t > \gamma + 1\}|$$
$$+ \|\hat{c}\| + \|\hat{d}\| + \beta \cdot |\{t : \hat{c}_t \geq D - \gamma\}| - (\hat{k} + \hat{l})D + (-A - \gamma)\beta. \quad (5.3.26b)$$

Case C'. $-B - 1 > \gamma$. Here we have

$$\mathrm{ymaj}_{\beta;\gamma+1}(Q_1) + \mathrm{ymaj}_{\beta;\gamma}(Q_2) = \|\hat{a}\| + \|\hat{b}\| + \beta \cdot |\{t : \hat{a}_t \geq D - \gamma - 1\}|$$
$$+ \|\hat{c}\| + \|\hat{d}\| + \beta \cdot |\{t : \hat{c}_t \geq D - \gamma\}| - (\hat{k} + \hat{l})D + (-A - \gamma)\beta + (-B - \gamma - 1)\beta.$$
$$(5.3.26c)$$

What concerns the right-hand side of (5.3.5), we have

$$\mathrm{ymaj}_{\beta;\gamma_2+1}(Q_1) + \mathrm{xmaj}_{\beta;\gamma_1+1}(Q_2) = \|\hat{a}\| + \|\hat{b}\| + \beta \cdot |\{t : \hat{b}_t > \gamma_2 + 1\}|$$
$$+ \|\hat{c}\| + \|\hat{d}\| + \beta \cdot |\{t : \hat{c}_t \geq \gamma_1 + 1\}| - (\hat{k} + \hat{l})D, \quad (5.3.27)$$

provided that $A + D < \gamma_1$ and $-B - 1 \leq \gamma_2$.

Now we are in the position to define Ψ. For any intersecting pair (P_1, P_2) of paths, $P_1 : \mathcal{A} \to \mathcal{E}$, $P_2 : \mathcal{B} \to \mathcal{F}$, perform the steps 1–4. The image of (P_1, P_2) under application of Ψ by definition is the resulting pair (Q_1, Q_2),

$$\Psi\big((P_1, P_2)\big) := (Q_1, Q_2).$$

It has already been shown that Ψ is a bijection. Considering all cases for γ, a combination of (5.3.7a,b,c), (5.3.17), (5.3.18a,b,c,e), respectively (5.3.24), (5.3.22), (5.3.26a,b,c), together with the remark that (5.3.18a,b,c,e) also hold if $\bar{a}, \bar{b}, \bar{c}, \bar{d}$ are replaced by $\hat{a}, \hat{b}, \hat{c}, \hat{d}$, respectively, by a little bit of manipulation finally shows, that

(5.3.2), and hence also (5.3.3) and (5.3.4), indeed hold. To establish (5.3.5) we have to combine (5.3.8a,b), (5.3.17), (5.3.19a), respectively (5.3.25), (5.3.22), (5.3.27).

Finally we turn to the statement in the Proposition concerning the last meeting point. Remember that in the second step I and J have been maximal with respect to (5.3.9) which had the meaning that (a_I, d_J) is the last meeting point between P_1 and P_2, where a_I and d_J refer to the array representations (5.3.6a,b) of P_1 and P_2. On the other hand, $(\hat{a} \mid \hat{b})$ and $(\hat{c} \mid \hat{d})$, defined via (5.3.10) and (5.3.20), were the array representations of Q_1 and Q_2, respectively. A short look at (5.3.10) and (5.3.20) confirms that Q_1 and Q_2 were constructed just in such a manner that (a_I, d_J) is the last meeting point of Q_1 and Q_2, too. Besides, the last $k - I + 1$ elements in the corresponding rows of (5.3.6a) and (5.3.10a) are identical, as well as the last $l - J$ elements in the corresponding rows of (5.3.6b) and (5.3.10b). Therefore the tails of P_1 and Q_1 and the tails of P_2 and Q_2 beginning from this last meeting point are identical.

The proof of the Proposition now is complete.

Again, we provide an example. Consider the paths $P_1^{\mathrm{ex}} : \mathcal{A}_{\mathrm{ex}} \to \mathcal{E}_{\mathrm{ex}}$ and $P_2^{\mathrm{ex}} : \mathcal{B}_{\mathrm{ex}} \to \mathcal{F}_{\mathrm{ex}}$ in Figure 7, $P_1^{\mathrm{ex}} = ((-1,0), 1121112122112212)$, $P_2^{\mathrm{ex}} = ((1,-2), 122212211$ $121121121)$, $\mathcal{A}_{\mathrm{ex}} = (-1,0)$, $\mathcal{E}_{\mathrm{ex}} = (8,7)$, $\mathcal{B}_{\mathrm{ex}} = (1,-2)$, $\mathcal{F}_{\mathrm{ex}} = (11,6)$. The array representations of $(P_1^{\mathrm{ex}}, P_2^{\mathrm{ex}})$ which are formed in the first step are

$$\begin{pmatrix} -1 \le & 1\ 4\ 5\ 7 & \le 7 & 1 \le & 2\ 3\ 6\ 8\ 10 & \le 10 \\ 1 \le & 1\ 2\ 4\ 6 & \le 7 & -1 \le & 1\ 3\ 4\ 5\ \ 6 & \le 6 \end{pmatrix}.$$

The integers I, J which are maximal with respect to (5.3.9) are given by $I = 4$ and $J = 3$. They correspond to the last meeting point $\mathcal{S}_{\mathrm{ex}} = (a_4^{\mathrm{ex}}, d_3^{\mathrm{ex}}) = (7,4)$ of P_1^{ex} and P_2^{ex}. Hence, by (5.3.10) this pair of arrays is mapped in the second step to the pair

$$\begin{pmatrix} 0 \le & 1\ 2\ 5\ 7 & \le 7 & 0 \le & 2\ 5\ 6\ 8\ 10 & \le 10 \\ 0 \le & 2\ 4\ 6 & \le 7 & 0 \le & 0\ 1\ 3\ 4\ 5\ \ 6 & \le 6 \end{pmatrix}.$$

Next (5.3.20) is applied to obtain from this pair the pair of arrays

$$\begin{pmatrix} 1 \le & 1\ 2\ 5\ 7 & \le 7 & -1 \le & 2\ 5\ 6\ 8\ 10 & \le 10 \\ -1 \le & -1\ 2\ 4\ 6 & \le 7 & 1 \le & 1\ 3\ 4\ 5\ \ 6 & \le 6 \end{pmatrix}.$$

These two arrays are the array representations for the paths Q_1^{ex} and Q_2^{ex} in Figure 7. Q_1^{ex} and Q_2^{ex} also meet in $\mathcal{S}_{\mathrm{ex}} = (7,4)$ for the last time, and the path portions after this last meeting point are identical with the corresponding portions of P_1^{ex} and P_2^{ex}.

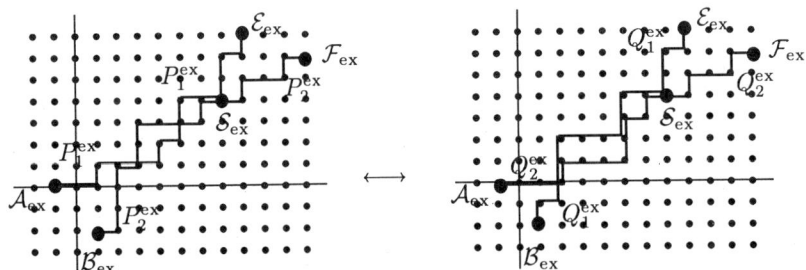

Figure 7

To illustrate (5.3.2), first choose $\gamma = 0$. By (4.0.1) we have $\text{ymaj}_{\beta;1}(P_1^{\text{ex}}) + \text{ymaj}_{\beta;0}(P_2^{\text{ex}}) = 34 + 3\beta + 53 + 5\beta = 87 + 8\beta$ and $\text{ymaj}_{\beta;1}(Q_1^{\text{ex}}) + \text{ymaj}_{\beta;0}(Q_2^{\text{ex}}) = 30 + 3\beta + 55 + 5\beta = 85 + 8\beta$ in accordance with (5.3.2). For $\gamma = -2$ by (4.0.1)–(4.0.3) we have $\text{ymaj}_{\beta;-1}(P_1^{\text{ex}}) + \text{ymaj}_{\beta;-2}(P_2^{\text{ex}}) = 34 + 4\beta + \beta + 53 + 5\beta = 87 + 10\beta$ and $\text{ymaj}_{\beta;-1}(Q_1^{\text{ex}}) + \text{ymaj}_{\beta;-2}(Q_2^{\text{ex}}) = 30 + 3\beta + 55 + 5\beta + 2\beta = 85 + 10\beta$ again in accordance with (5.3.2). It is left to the reader to try other cases. $\square$

REMARK. We promised to try to give an explanation why nothing reasonable can be expected unless the starting points $\mathcal{A}$ and $\mathcal{B}$ of the paths P_1 and P_2 lie on a line parallel to $x + y = 0$. Remember that the aim has been to find some correspondence between intersecting pairs (P_1, P_2) of paths $P_1 : \mathcal{A} \to \mathcal{E}$, $P_2 : \mathcal{B} \to \mathcal{F}$, and arbitrary pairs (Q_1, Q_2) of paths $Q_1 : \mathcal{B} \to \mathcal{E}$, $Q_2 : \mathcal{A} \to \mathcal{F}$, such that the tails of P_1 and Q_1, and the tails of P_2 and Q_2 after the last meeting point between P_1 and P_2, which in addition is the same for Q_1 and Q_2, are identical. We are basically counting by the major index. Consider any North-East corner in P_1, say, which comes after this last meeting point. This corner contributes the number of steps from $\mathcal{A}$ to this corner to the major index or strange major index. If the desired correspondence would obey the above described property, then this North-East corner would be also a North-East corner for Q_1. But if $\mathcal{B}$, the starting point of Q_1, did not lie on the same line parallel to $x + y = 0$ as $\mathcal{A}$ does, then the number of steps from the starting point of Q_1 to this corner would be completely different. Of course the same is true with North-East corners of P_2 and Q_2. Since the number of North-East corners which occur after this last meeting point is not predictable, the difference which would be caused if $\mathcal{A}$ and $\mathcal{B}$ would not lie on a line parallel to $x + y = 0$ could not be handled properly.

5.4. Correspondences between tableaux and nonintersecting lattice paths. Here we explain the connection between the major and strange major counting of nonintersecting lattice paths and the computation of generating functions for tableaux. The ideas which are used are the celebrated Knuth correspondences [24], one of Burge's [6, p. 22] modifications of it, the geometric interpretations of Knuth's and Burge's correspondences due to Viennot [42] and Desainte-Catherine and Viennot [10] and a refinement of Choi and Gouyou–Beauchamps [7]. Though our presentation sometimes differs from the presentations in the cited papers, in fact all the statements of this subsection are, at least implicitly, in these papers. Therefore the proofs for the following Propositions will be kept short and cursory. What could be called new is the observation that the norm of tableaux carries over into the major index or strange major index of nonintersecting lattice paths.

The main tools of the Knuth correspondences and its variants are algorithms which insert a positive integer into a tableaux or delete an integer from a tableaux, respectively.

We start with the algorithms ROW-INSERT and ROW-DELETE. Given a tableau τ and a positive integer x the algorithm ROW-INSERT inserts x into τ by the following procedure.

Algorithm ROW-INSERT [24, p. 712].

(1) Compare x with the elements of the first row of τ. While reading from left

to right, replace the first element which is larger than x by x. If there is no larger element then add x as last element to the first row.

(2) If an element was "bumped" in the previous step, insert it into the second row following the same rule. Repeat this with the subsequent rows until the element to be inserted can be placed at the end of a (possibly empty) row.

(3) Denote the tableaux which results from this inserting procedure by ROW-INSERT(τ, x).

Given the output $\bar{\tau} = \text{ROW-INSERT}(\tau, x)$ of such an inserting procedure, we are able to recover τ and x if we know the number r of the row of $\bar{\tau}$ where ROW-INSERT stopped. This is accomplished by the algorithm ROW-DELETE. Let $\bar{\tau}$ be a tableau and r a positive integer.

Algorithm ROW-DELETE [24, p. 713].

(1) Take the last element of the r-th row of $\bar{\tau}$. Use it to bump and replace the right-most element of the $(r-1)$-th row that is smaller.

(2) Take the bumped element and repeat this procedure in the $(r-2)$-th row, etc.

(3) The element which is finally bumped in the first row is denoted by x, the remaining tableau by τ. The output of ROW-DELETE($\bar{\tau}, r$) is (τ, x).

Dually, there are column insertion and column deletion. The difference is that now we insert into the columns and that everywhere strict order is replaced by weak order.

Algorithm COLUMN-INSERT [24, p. 720].

(1) Compare x with the elements of the first column of τ. While reading from top to bottom, replace the first element which is greater or equal to x by x. If there is no such element then add x as last element to the first column.

(2) If an element was "bumped" in the previous step, insert it into the second column following the same rule. Repeat this with the subsequent columns until the element to be inserted can be placed at the bottom of a (possibly empty) column.

(3) Denote the tableaux which results from this inserting procedure by COLUMN-INSERT(τ, x).

Also with column-insertion, given the number r of the column of the tableau $\bar{\tau} = \text{COLUMN-INSERT}(\tau, x)$ where COLUMN-INSERT stopped, we are able to recover τ and x, this time using COLUMN-DELETE.

Algorithm COLUMN-DELETE [24, p. 720].

(1) Take the last element of the r-th column of $\bar{\tau}$. Use it to bump and replace the down-most element of the $(r-1)$-th column that is smaller or equal.

(2) Take the bumped element and repeat this procedure in the $(r-2)$-th column, etc.

(3) The element which is finally bumped in the first column is denoted by x, the remaining tableau by τ. The output of COLUMN-DELETE($\bar{\tau}, r$) is (τ, x).

We begin with two correspondences for pairs of tableaux.

Proposition 28. *There exists a bijection Δ_1 between pairs (τ_1, τ_2) of tableaux of identical shape, with at most r columns and with parts between 1 and n, and* families $\mathfrak{P} = (P_1, \ldots, P_r)$ of nonintersecting lattice paths, $P_i : (i-1, -i+1) \to (n+i-1, n-i+1)$, $i = 1, \ldots, r$, such that

$$\mathrm{ymaj}_{1,0}\, \Delta_1(\tau) = n(\tau_1) + n(\tau_2). \tag{5.4.1}$$

(The extension of the strange major index ymaj to families of lattice paths is introduced in (4.0.8).)

PROOF. First we give a bijection between pairs (τ_1, τ_2) of tableaux of identical shape and two-line arrays $(u \mid v)$,

$$\begin{pmatrix} u_1 & u_2 & \ldots & u_N \\ v_1 & v_2 & \ldots & v_N \end{pmatrix} \tag{5.4.2}$$

of positive integers which satisfy

$$\begin{aligned} i \le j \quad \Rightarrow \quad &\text{either} \quad u_i < u_j \\ &\text{or} \quad u_i = u_j \text{ and } v_i \ge v_j. \end{aligned} \tag{5.4.3}$$

As an example of an array satisfying (5.4.3) consider the two-line array in Figure 8. Given an array $(u \mid v)$ satisfying (5.4.3) we form a sequence $(\tau_1^{(i)}, \tau_2^{(i)})_{i=0,\ldots,N}$ of pairs of tableaux. We start with $\tau_1^{(0)} = \tau_2^{(0)} = \emptyset$ ($\emptyset$ denotes the empty tableau). Then for $i = 1, \ldots, N$ we set $\tau_1^{(i)} = \mathrm{COLUMN\text{-}INSERT}(\tau_1^{(i-1)}, v_i)$, and, if r is the number of the column of $\tau_1(i)$ where COLUMN-INSERT stopped while inserting v_i into $\tau_1^{(i-1)}$, we place u_i at the bottom of the r-th column of $\tau_2^{(i-1)}$ thus obtaining $\tau_2(i)$. Finally we set $(\tau_1, \tau_2) = (\tau_1^{(N)}, \tau_2^{(N)})$. In Figure 8 the reader finds an example of a two-line array and a pair of tableaux which correspond to each other by this procedure.

It can be seen that this mapping is a bijection between pairs (τ_1, τ_2) of tableaux of the same shape, with parts between 1 and n, and with exactly r_0 columns, and two-line arrays $(u \mid v)$ satisfying (5.4.3) which consist of integers between 1 and n and where the longest (weakly) decreasing subsequence of v, that is a sequence

$$v_{i_1} \ge v_{i_2} \ge \cdots \ge v_{i_j} \quad \text{with} \quad i_1 \le i_2 \le \cdots \le i_j, \quad \text{with } j \text{ maximal}, \tag{5.4.4}$$

has length r_0. The non-obvious part is the equality between the number of columns of the tableaux and the maximal length of a decreasing subsequence of the bottom row of the two-line array. Since this statement is only implicit in existing papers [24, 6, sec. 3] we sketch a proof at the end of the proof of this Proposition. For later use we denote the mapping from a two-line array to a pair of tableaux by K_C. In the tableaux in Figure 8 the length of the first row is 3, and longest decreasing subsequences of the bottom row of the two-line array in Figure 8 for instance are $(4, 4, 2)$, $(6, 5, 3)$, $(6, 6, 6)$, all being of length 3.

EXAMPLE with $n = 7$, $r_0 = 3$, $r = 4$

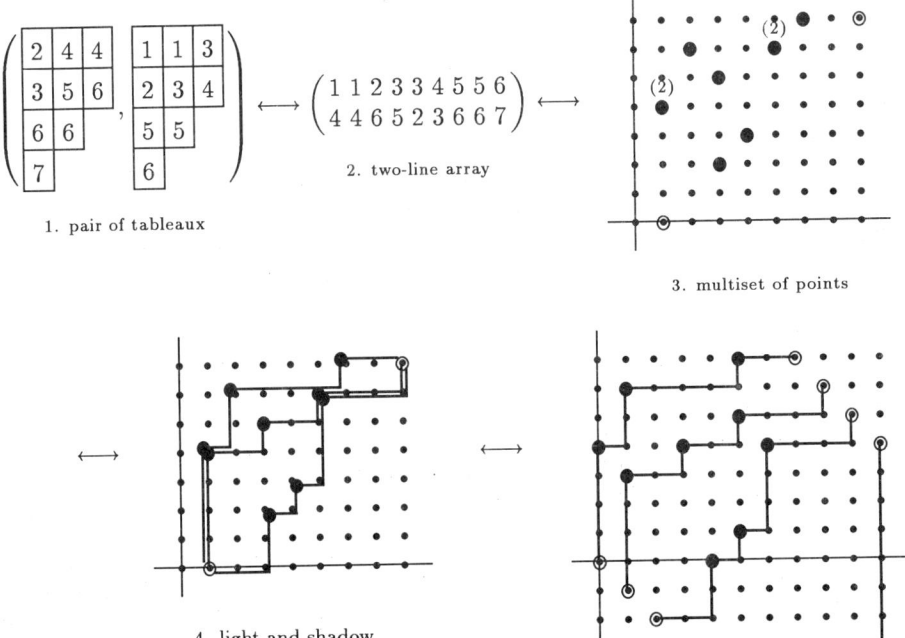

1. pair of tableaux

2. two-line array

3. multiset of points

4. light and shadow

5. family of nonintersecting lattice paths

Figure 8

Next the array $(u \mid v)$ is viewed geometrically by interpreting (u_i, v_i), $i = 1, \ldots, N$, as points (with multiplicity) in the integer lattice $\mathbb{Z}^2$. Besides we add the points $(1, 0)$ and $(n + 1, n)$ to this multiset of points. In the third picture of Figure 8 these two points are indicated as circles. The symbol (2) being located above the points $(1, 4)$ and $(5, 6)$ indicates that these points have multiplicity 2.

Now Viennot's *light and shadow* procedure [42, 10] is applied. We suppose that there is a light being located in the North-West of the multiset of points. The *shadow* of a point (x, y) is defined to be the set of points $(x', y') \in \mathbb{R}^2$ with $x \leq x' \leq n + 1$ and $0 \leq y' \leq y$. We consider the (North-West) *border* of the union of the shadows of all the points of the multiset. This border is a lattice path from $(1, 0)$ to $(n + 1, n)$. The North-East corners of this path are elements of the given multiset. Now we remove one copy of each of these North-East corners from the multiset. Then the light and shadow procedure is repeated, as long as there are points left. A moment's thought shows that if in a two-line array $(u \mid v)$ the maximal length of a (weakly) decreasing subsequence of v is r_0, then by the light and shadow procedure we obtain exactly r_0

lattice paths. By construction, these lattice paths may meet but do not cross each other. Item 4 in Figure 8 displays the lattice paths which in our example are obtained by this procedure.

Next we shift the i-th path in the direction $(i-2, -i+1)$, $i = 1, \ldots, r_0$, thus obtaining a family $(P_1, \ldots, P_{r_0})$ of *nonintersecting* lattice paths, $P_i : (i-1, -i+1) \to (n+i-1, n-i+1)$, $i = 1, \ldots, r_0$. Finally we add the paths $P_{r_0+1}, \ldots, P_r$, $P_i : (i-1, -i+1) \to (n+i-1, n-i+1)$, $i = r_0+1, \ldots, r$, all of which start with n horizontal steps followed by n vertical steps. In our example the resulting family of nonintersecting lattice paths is item 5 of Figure 8.

To verify (5.4.1), observe that in the correspondence between pairs of tableaux and two-line arrays the elements are preserved. Therefore the sum $n(\tau_1) + n(\tau_2)$ of the norms of the original tableaux equals the sum of all the coordinates of the North-East corners of the paths which are obtained by the light and shadow procedure (item 4 in Figure 8). The contribution of a North-East corner (x, y) to $\mathrm{ymaj}_{1,0}(P)$ of a path P which starts in $(1, 0)$ is just the number of steps from $(1, 0)$ to (x, y) plus 1, that is $(x + y - 1) + 1 = x + y$. Hence $n(\tau_1) + n(\tau_2)$ also equals $\sum \mathrm{ymaj}_{1,0} P$, where the sum is over all paths P which are obtained by the light and shadow procedure. The final shifting and adding of paths converts this sum into $\sum_{i=1}^{r} \mathrm{ymaj}_{1,-i+1} P_i$, P_i being the i-th path of the resulting family of nonintersecting lattice paths. Comparison with (4.0.8) yields (5.4.1). For the pair of tableaux $(\tau_1^{\mathrm{ex}}, \tau_2^{\mathrm{ex}})$, say, in Figure 8 we have $n(\tau_1^{\mathrm{ex}}) + n(\tau_2^{\mathrm{ex}}) = 43 + 30 = 73$. For the resulting family $\mathfrak{P}^{\mathrm{ex}} = (P_1^{\mathrm{ex}}, P_2^{\mathrm{ex}}, P_3^{\mathrm{ex}}, P_4^{\mathrm{ex}})$ of nonintersecting lattice paths we have

$$\mathrm{ymaj}_{1,0}(\mathfrak{P}^{\mathrm{ex}}) = \mathrm{ymaj}_{1,0}(P_1^{\mathrm{ex}}) + \mathrm{ymaj}_{1,-1}(P_2^{\mathrm{ex}}) + \mathrm{ymaj}_{1,-2}(P_3^{\mathrm{ex}}) + \mathrm{ymaj}_{1,-3}(P_4^{\mathrm{ex}})$$
$$= 26 + 24 + 23 + 0 = 73,$$

in accordance with (5.4.1).

As promised, at the end we sketch a proof that the mapping K_C that was defined above and maps a two-line array $(u \mid v)$ to a pair of tableaux (τ_1, τ_2) has the property that the longest decreasing subsequence of v (see (5.4.4)) has length r_0 if and only if the number of columns in τ_1 and τ_2 is r_0. This can be seen in several different ways. We decide to prove it by means of Viennot's [42, 38, sec. 3.8] geometric interpretation of the Knuth correspondence since we shall also need it at another place. Actually, in [42, 38, sec. 3.8] only the Robinson–Schensted correspondence is considered. But it is easy to see that everything carries over to the Knuth correspondence.

Assuming that the reader is already familiar with either [42] or [38, sec. 3.8], we briefly recall Viennot's construction. Let $(u \mid v)$ be a two-line array of the form (5.4.2) such that

$$i \leq j \quad \Rightarrow \quad \text{either} \quad u_i < u_j$$
$$\text{or} \quad u_i = u_j \text{ and } v_i \leq v_j. \qquad (5.4.5)$$

Note that this differs from (5.4.3) in the behaviour of v_i's beneath equal u_i's. Suppose that with this two-line array we do the same as under the mapping K_C, but using ROW-INSERT instead of COLUMN-INSERT. Let us denote this mapping by K_R

and the resulting pair of tableaux by (P_R, Q_R). Again $(u \mid v)$ is viewed as a multiset of points in $\mathbb{Z}^2$, only that now $(1, 0)$ and $(n + 1, n)$ are not added. Next we assume a light to be located in the South-West of the multiset of points. By light and shadow we obtain a set of lattice paths, each having exactly one infinite horizontal and one infinite vertical ray. Viennot's theorem says that we can read off the first row of P_R by the y-coordinates of the horizontal rays. Likewise, we can read off the first row of Q_R by the x-coordinates of the vertical rays. In order to get the second rows, we take the "skeletons" of the paths (see [42, 38, sec. 3.8]), apply light and shadow and read the y-coordinates of the horizontal rays and the x-coordinates of the vertical rays, as before. This procedure is iterated to obtain the remaining rows of P_R and Q_R. Simplifying, we say '*with a light in the South-West the rows of P_R can be read off from the right and the rows of Q_R can be read off from the top*'. This terminology clearly is motivated by the picture of the construction.

So far we considered ROW-INSERTion. What we need is the analogue of Viennot's construction for COLUMN-INSERTion. Let $(u \mid v)$ be a two-line array, ordered according to (5.4.3). Let (P_C, Q_C) be the pair of tableaux that results from $(u \mid v)$ by applying K_C. Again interpret $(u \mid v)$ as a multiset of points in $\mathbb{Z}^2$. Our claim is that *with a light located in the South-East the rows of P_C can be read off from the left, and that with a light in the North-West the rows of Q_C can be read off from the bottom*. This is the geometric interpretation of the COLUMN-Knuth correspondence K_C. It does not seem to have been recorded before. It is easily checked for our example in Figure 8 (items 1 and 4).

First we turn to the reading of P_C. It is known that if the bottom row v of the two-line array by repeated COLUMN-INSERT is mapped to a tableau we would obtain the same as when mapping $v^r := v_N v_{N-1} \ldots v_1$, the reversion of v, to a tableau by using repeated ROW-INSERT. This is the contents of e.g. [38, Theorem 4.8.8; the equation should read $T(\tilde{\pi}^r) = T'(\tilde{\pi})$, where T' denotes application of $R - S - K'$, the repeated COLUMN-INSERTion]. Hence, if we apply K_R to $(\tilde{u} \mid v^r)$, where v^r is as before and $\tilde{u}$ is defined by $\tilde{u}_i = U - u_{N-i+1}$, $i = 1, 2, \ldots, N$, with U larger than all u_i's, the resulting tableau P_R would be the same as the tableau P_C which is obtained by applying K_C to $(u \mid v)$. Note that geometrically the multiset of points corresponding to $(\tilde{u} \mid v^r)$ results from the multiset of points corresponding to $(u \mid v)$ by a reflection in the vertical line $x = U/2$. Also, if $(u \mid v)$ is ordered according to (5.4.3) then $(\tilde{u} \mid v^r)$ is ordered according to (5.4.5). This guarantees that application of K_R to $(\tilde{u} \mid v^r)$ is indeed allowed. But we know how to read the rows of P_R: With a light in the South-West of $(\tilde{u} \mid v^r)$ (viewed as multiset of points) the rows of P_R are read from the right. Since P_R (corresponding to $(\tilde{u} \mid v^r)$) is the same as P_C (corresponding to $(u \mid v)$), and since $(u \mid v)$ results from $(\tilde{u} \mid v^r)$ by a reflection in the vertical line $x = U/2$, we know how to read P_C: The light in the South-West under the reflection in the vertical line becomes a light in the South-East, instead of reading from the right after the reflection we are reading from the left. This proves the claim about the reading of P_C.

In order to prove the claim about the reading of Q_C, we recall another standard property of the COLUMN-Knuth correspondence K_C: If $(u \mid v)$ by K_C is mapped to (P_C, Q_C) then the transpose $(u \mid v)^t$ by K_C is mapped to (Q_C, P_C). The transpose

$(u \mid v)^t$ is defined by interchanging the top and bottom row of $(u \mid v)$ and then reordering the pairs $\binom{v_i}{u_i}$ according to (5.4.3). A proof of this fact is sketched in [6, p. 21]. Note that, geometrically, the multiset of points corresponding to $(u \mid v)^t$ results from the multiset of points corresponding to $(u \mid v)$ by a reflection in the diagonal $x = y$. We have already proved that the first tableau, P_C, that is obtained by K_C is read with a light in the South-East from the left. Hence, Q_C can be read off $(u \mid v)^t$ with a light in the South-East from the left. After reflecting back we see that Q_C can be read off $(u \mid v)$ with a light in the North-West from the bottom.

Now we are prepared to attack our original problem. Let $(u \mid v)$ be a two-line array ordered according to (5.4.3). It is easy to see that the longest (weakly) decreasing subsequence of v has length r_0 if and only if with a light in the South-East (or in the North-West) light and shadow would produce exactly r_0 lattice paths. But, by our geometric interpretation of K_C, this means that in the pair of tableaux (τ_1, τ_2) that corresponds to $(u \mid v)$ under application of K_C the first rows of τ_1 and τ_2 have length r_0. $\quad \square$

Quite analogously, we obtain a result for pairs of tableaux with odd entries.

Proposition 29. *There exists a bijection Δ_2 between pairs (τ_1, τ_2) of tableaux of identical shape, with at most r columns and with only odd parts which lie between 1 and $2n - 1$, and families $\mathfrak{P} = (P_1, \ldots, P_r)$ of nonintersecting lattice paths, $P_i :$ $(2i - 2, -2i + 2) \to (2n + 2i - 2, 2n - 2i + 2)$, $i = 1, \ldots, r$, where the North-East corners of the paths have odd coordinates, such that*

$$\operatorname{maj} \Delta_2(\tau) = n(\tau_1) + n(\tau_2). \qquad (5.4.6)$$

(The extension of the major index to families of lattice paths is introduced in (3.0.1).)

PROOF. We do almost the same as in the previous proof. First, as described above, we map a pair (τ_1, τ_2) of tableaux of the type under consideration onto its corresponding two-line array. This array is interpreted geometrically as multiset of points in $\mathbb{Z}^2$ as before, but now we add the points $(0, 0)$ and $(2n, 2n)$ as the future starting and final points of the paths which are subsequently constructed by the light and shadow procedure. Next the i-th path is shifted in the direction $(2i - 2, -2i + 2)$. If r_0 is the number of columns in τ_1 (and hence also in τ_2), then by now we obtained r_0 pairwise nonintersecting paths, $P_1, \ldots, P_{r_0}$. Finally we add the paths $P_{r_0+1}, \ldots, P_r$, $P_i : (2i - 2, -2i + 2) \to (2n + 2i - 2, 2n - 2i + 2)$, $i = r_0 + 1, \ldots, r$, all of which start with $2n$ horizontal steps followed by $2n$ vertical steps.

The asserted properties of this bijection are readily verified. $\quad \square$

Now we turn to correspondences for single tableaux. They are based on Burge's correspondence [6, p. 22].

Proposition 30. *There exists a bijection Δ_3 between tableaux τ with rows of even lengths, with at most $2r$ columns, and with parts between 1 and n, and families $\mathfrak{P} = (P_1, \ldots, P_r)$ of nonintersecting lattice paths which lie below the line $x = y$, $P_i : (i - 1, -i + 1) \to (n + i, n + 2 - i)$, $i = 1, \ldots, r$, such that*

$$\operatorname{maj} \Delta_3(\tau) = n(\tau). \qquad (5.4.7)$$

Moreover Δ_3 has the property that if the first row of τ has length $2r$ and if the last entry in the first row of τ is h then the r-th path of the family $\Delta_3(\tau)$ starts with h horizontal steps followed by a vertical step. If the length of the first row of τ is less than $2r$ then the r-th path of $\Delta_3(\tau)$ starts with $n+1$ horizontal steps (of course followed by $n+1$ vertical steps).

PROOF. We follow Desainte–Catherine and Viennot [10], but we present the tableaux in English notation.

First we set up a bijection between tableaux τ with even rows and two-line arrays $(u \mid v)$ of the form (5.4.2) which satisfy (5.4.3), and in addition satisfy

$$u_i \geq v_i \quad \text{for } i = 1, 2 \ldots, N. \tag{5.4.8}$$

Given such an array $(u \mid v)$ we form the sequence $(\tau^{(i)})_{i=0,\ldots,N}$ of tableaux, where $\tau^{(0)}$ is the empty tableau and $\tau^{(i)}$ is obtained from $\tau^{(i-1)}$ and the pair (u_i, v_i) in the following way. Insert v_i into $\tau^{(i-1)}$ using COLUMN-INSERT. Then add u_i at the end of the row of COLUMN-INSERT($\tau^{(i-1)}, v_i$) where COLUMN-INSERT stopped. The resulting tableau is $\tau^{(i)}$, $i = 1, \ldots, N$. Finally we set $\tau = \tau^{(N)}$. The tableau and the two-line array in Figure 9 correspond to each other by this procedure.

It can be seen that this mapping is a bijection between tableaux with even rows, with exactly $2r_0$ columns, and with parts between 1 and n, and two-line arrays (5.4.2) satisfying (5.4.3) and (5.4.8) where the longest (weakly) decreasing subsequence of v, i.e. a subsequence satisfying (5.4.4), equals r_0. Again, the non-obvious part is the correspondence between the number of columns of the tableau and the maximal length of a decreasing subsequence of the bottom row of the two-line array. Also this statement has not been proved explicitly, therefore we sketch a proof at the end of the proof of this Proposition. For later use we denote the mapping from a two-line array to a tableau with even rows by K_E. In the tableau in Figure 9 the length of the first row is 6, and the longest decreasing subsequences of the two-line array in Figure 9 are $(3, 3, 1)$ and $(3, 3, 2)$, both being of length 3.

Again the array (u, v) is interpreted as multiset of points (u_i, v_i) in $\mathbb{Z}^2$. Because of (5.4.8) all the points are located below or on the diagonal $x = y$. This time we add the points $(0, 0)$ and $(n + 1, n + 1)$. Once more the light and shadow procedure is performed, yielding a set of paths, all of which running from $(0, 0)$ to $(n + 1, n + 1)$, which do not cross each other and lie below $x = y$. If the maximal length of a (weakly) decreasing subsequence of v is r_0, then by the light and shadow procedure we obtain exactly r_0 paths. Shifting the i-th path in the direction $(i - 1, -i + 1)$ gives a family $(P_1, \ldots, P_{r_0})$ of nonintersecting lattice paths which lie below $x = y$, $P_i : (i - 1, -i + 1) \to (n + i, n - i + 2)$. Finally the paths $P_{r_0+1}, \ldots, P_r, P_i : (i - 1, -i + 1) \to (n + i, n - i + 2)$, are added, each of them starting with $n + 1$ horizontal steps followed by $n + 1$ vertical steps. This sets up the desired bijection. Relation (5.4.7) can be readily verified using similar arguments to those in the proof of Proposition 28. For our tableau τ^{ex}, say, in Figure 9 we have $n(\tau^{\text{ex}}) = 33$, while for the corresponding family $\mathfrak{P}^{\text{ex}} = (P_1^{\text{ex}}, P_2^{\text{ex}}, P_3^{\text{ex}})$ of nonintersecting lattice paths in Figure 9 we have $\text{maj } P_1^{\text{ex}} + \text{maj } P_2^{\text{ex}} + \text{maj } P_3^{\text{ex}} = 17 + 6 + 10 = 33$, in accordance with (5.4.7).

EXAMPLE with $n = 5$, $r_0 = r = 3$, $h = 3$

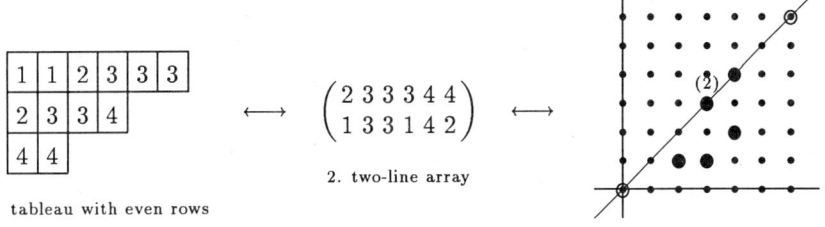

1. tableau with even rows

2. two-line array

$$\begin{pmatrix} 2\ 3\ 3\ 3\ 4\ 4 \\ 1\ 3\ 3\ 1\ 4\ 2 \end{pmatrix}$$

3. multiset of points

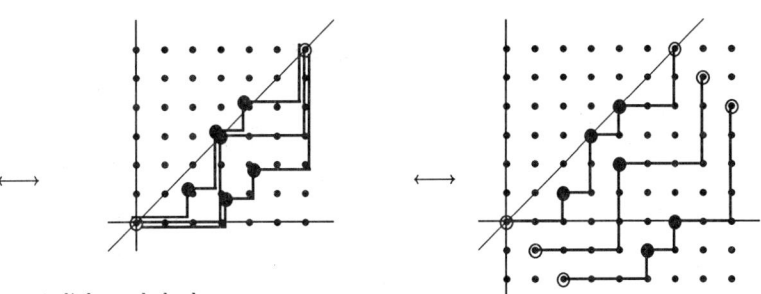

4. light and shadow

5. family of nonintersecting lattice paths

Figure 9

Finally we have to confirm the last two assertments of the Proposition. In case that the tableau has less than $2r$ columns, by the light and shadow procedure less than r paths would be obtained. Hence P_r would be one of the added paths starting with $n + 1$ horizontal steps followed by $n + 1$ vertical steps. This proves the last assertment. To establish the next-to-last assertment, observe that if P_r starts with h steps followed by a vertical step there is a $\bar{v}$ such that the point $(h, \bar{v})$ is the first North-East corner of the r-th path of the family of paths that is obtained by the light and shadow procedure. So the corresponding two-line array looks like

$$\begin{pmatrix} u_1 & \ldots & u_{k-1} & h & u_{k+1} & \ldots & u_N \\ v_1 & \ldots & v_{k-1} & \bar{v} & v_{k+1} & \ldots & v_N \end{pmatrix}. \tag{5.4.9}$$

Besides, $\bar{v}$ is the last element in a decreasing subsequence $(v_{i_1}, \ldots, v_{i_{r-1}}, \bar{v})$ of v of maximal length. Also, all the maximal decreasing subsequences of the sequence

$$(v_1, \ldots, v_{k-1}) \tag{5.4.10}$$

have length $r - 1$. Now recall that in order to obtain the tableau τ that corresponds to the array (5.4.9), we have to form the sequence $\tau^{(0)}, \tau^{(1)}, \ldots, \tau^{(N)}$ of tableaux by using the array (5.4.9) and this insertion procedure. Since all the maximal decreasing subsequences of the sequence (5.4.10) have length $r - 1$, the length of the first row of $\tau^{(k-1)}$ is $2r-2$. Next we have to COLUMN-INSERT $\bar{v}$ into $\tau^{(k-1)}$. Since $\bar{v}$ belongs to a maximal decreasing subsequence of length r in the bottom row of (5.4.9), COLUMN-INSERT$(\tau^{(k-1)}, \bar{v})$ will stop at the end of the first row of $\tau^{(k-1)}$. Subsequently, h is added at the end of the first row of COLUMN-INSERT$(\tau^{(k-1)}, \bar{v})$ to obtain $\tau^{(k)}$. Everything will be established if we show that h cannot be bumped during the remaining insertions. But this is obvious, because if h is bumped a new column would be created. Thus the resulting tableau τ would consist of more than $2r$ columns, a contradiction to our assumption.

As promised, at the end we sketch a proof that the mapping K_E that was defined above and maps a two-line array $(u \mid v)$ which satisfies (5.4.3) and (5.4.8) to a tableau τ with even rows has the property that the longest decreasing subsequence of v (see (5.4.4)) has length r_0 if and only if the number of columns of τ is $2r_0$. What we need is the following relation between the maps K_E and K_C (the latter appeared in the proof of Proposition 28) which was stated in [6, sec. 3] but was not proved there: There holds the relation

$$K_C\big((u \mid v) \cup (v \mid u)\big) = \big(K_E((u \mid v)), K_E((u \mid v))\big). \tag{5.4.11}$$

Here, the symbol $(u \mid v) \cup (v \mid u)$ denotes the "union" of $(u \mid v)$ and $(v \mid u)$. This is formed by concatenating $(u \mid v)$ and $(v \mid u)$ and reordering the concatenation according to (5.4.3). For example, the union of $\left(\begin{smallmatrix} 2 & 3 & 3 & 3 \\ 1 & 3 & 3 & 1 \end{smallmatrix}\right)$ and $\left(\begin{smallmatrix} 1 & 3 & 3 & 1 \\ 2 & 3 & 3 & 3 \end{smallmatrix}\right)$ is $\left(\begin{smallmatrix} 1 & 1 & 2 & 3 & 3 & 3 & 3 \\ 3 & 2 & 1 & 3 & 3 & 3 & 1 \end{smallmatrix}\right)$. Note that, when viewing $(u \mid v)$ and $(v \mid u)$ as multisets of points in $\mathbb{Z}^2$, the union $(u \mid v) \cup (v \mid u)$, as a multiset of points, indeed equals the multiset union of $(u \mid v)$ and $(v \mid u)$. More precisely, $(u \mid v) \cup (v \mid u)$ is the union of the multiset of points corresponding to $(u \mid v)$ and its reflection in the diagonal $x = y$. Thus $(u \mid v) \cup (v \mid u)$, as multiset of points, is symmetrical and all pairs of the form $\left(\begin{smallmatrix} w \\ w \end{smallmatrix}\right)$ occur in $(u \mid v) \cup (v \mid u)$ with an even multiplicity.

If we suppose that the relation (5.4.11) between the maps K_C and K_E holds then the above claim is easily proved: Obviously the maximal length of a decreasing subsequence of v, the bottom row of $(u \mid v)$, is r_0 if and only if the maximal length of a decreasing subsequence of the bottom row of the union $(u \mid v) \cup (v \mid u)$ is $2r_0$. (This is best seen geometrically.) On the other hand, in the proof of Proposition 28 we have shown that the maximal length of a decreasing subsequence of $(u \mid v) \cup (v \mid u)$ has length $2r_0$ if and only if the tableaux that result from $(u \mid v) \cup (v \mid u)$ under application of K_C have exactly $2r_0$ columns. Thus, by (5.4.11) our claim is established.

What remains is to prove (5.4.11). Let $(u \mid v)$ be a two-line array satisfying (5.4.3) and (5.4.8), and let τ be the tableau that results from $(u \mid v)$ under application of K_E. Since $(u \mid v) \cup (v \mid u)$ as a multiset of points is symmetric, application of K_C to $(u \mid v) \cup (v \mid u)$ yields a pair of identical tableaux. This is a standard property of K_C (cf. [6, p. 21]). Therefore we only have to prove that the *first* tableau equals $\tau = K_E((u \mid v))$. It is also true that application of K_C to $(u \mid v) \cup (v \mid u)$ yields a pair of (identical) tableaux with only *even* rows since pairs of the form $\left(\begin{smallmatrix} w \\ w \end{smallmatrix}\right)$ always occur with an even multiplicity in $(u \mid v) \cup (v \mid u)$. Again this is standard (cf. [6, p. 21]).

We perform an induction on the length N of the two-line array $(u \mid v)$. The relation (5.4.11) is trivially true for $N = 0$. Let $\bar{u}$ denote the sequence $u_1 u_2 \ldots u_{N-1}$, and similarly for $\bar{v}$. Suppose that (5.4.11) has already been proved for all two-line arrays of length $N - 1$. Using the geometric interpretation of K_C that was developed in the proof of Proposition 28, this means that the tableau $\bar{\tau}$ that results from $(\bar{u} \mid \bar{v})$ by application of K_E can be read off $(\bar{u} \mid \bar{v}) \cup (\bar{v} \mid \bar{u})$ with a light in the South-East from the left. To obtain τ we have to insert the pair (u_N, v_N) into the tableau $\bar{\tau}$ in the way described above. That is, first v_N is COLUMN-INSERTed into $\bar{\tau}$ and then u_N is placed at the end of the row where the algorithm stopped. Geometrically, COLUMN-INSERTion of v_N into $\bar{\tau}$ means to add the point (u_N, v_N) to the multiset of points $(\bar{u} \mid \bar{v}) \cup (\bar{v} \mid \bar{u})$ and with a light in the South-East read from the left. Now u_N is placed at the end of the row where the COLUMN-INSERTion stopped. Thus τ is obtained. We have to show that this is the same as reading the multiset of points corresponding to $(u \mid v) \cup (v \mid u)$ with a light in the South-East from the left. The multiset of points corresponding to $(u \mid v) \cup (v \mid u)$ is obtained from the multiset of points corresponding to $(\bar{u} \mid \bar{v}) \cup (\bar{v} \mid \bar{u}) \cup (u_N \mid v_N)$ by adding the point (v_N, u_N). It is easy to see that the effect that is caused by the addition of (v_N, u_N) when reading with a light in the South-East from the left will be the placement of u_N at the end of a row. This is due to the fact that, because of the order (5.4.3), the point (v_N, u_N) is among the highest points of $(u \mid v) \cup (v \mid u)$, viewed as a multiset of points, and among all those it is the left-most. Now, the tableau, say τ', that is read off $(\bar{u} \mid \bar{v}) \cup (\bar{v} \mid \bar{u}) \cup (u_N \mid v_N)$ is a tableau whose shape differs from the shape of $\bar{\tau}$, a tableau with only even rows, just in one cell. Hence, τ' has exactly one row of odd length. And τ' differs from the reading of $(u \mid v) \cup (v \mid u)$ by the placement of u_N at the end of a row of τ'. The result of the placement of u_N in τ' will again be a tableau with only even rows since, as was remarked above, pairs of the form $\binom{w}{w}$ occur with an even multiplicity in $(u \mid v) \cup (v \mid u)$. But because there is exactly one odd row in τ' this must be the row where u_N is added. Besides this is the row where the COLUMN-INSERTion of v_N into $\bar{\tau}$ stopped. Thus the reading of $(u \mid v) \cup (v \mid u)$ and the tableau τ agree. $\quad\square$

If we apply the same arguing to tableaux with only odd entries we obtain an analogous result.

Proposition 31. *There exists a bijection Δ_4 between tableaux τ with rows of even lengths, with at most $2r$ columns, and with only odd parts which lie between 1 and $2n - 1$, and families $\mathfrak{P} = (P_1, \ldots, P_r)$ of nonintersecting lattice paths which lie below the line $x = y$, $P_i : (2i - 2, -2i + 2) \to (2n + 2i - 2, 2n - 2i + 2)$, $i = 1, \ldots, r$, where the North-East corners of the paths have odd coordinates, such that*

$$\operatorname{maj} \Delta_4(\tau) = n(\tau). \tag{5.4.12}$$

Moreover Δ_4 has the property that if the first row of τ has the length $2r$ and if the last entry in the first row of τ is $2h + 1$ then the r-th path of the family $\Delta_4(\tau)$ starts with $2h + 1$ horizontal steps followed by a vertical step. If the length of the first row of τ is less than $2r$ then the r-th path of $\Delta_4(\tau)$ starts with $2n$ horizontal steps (of course followed by $2n$ vertical steps). $\quad\square$

Using the idea of Choi and Gouyou–Beauchamps, Proposition 30 can be extended to tableaux with a prescribed number of odd rows.

Proposition 32. *There exists a bijection Δ_5 between tableaux τ with p odd rows, with at most $2r$ columns, and with parts between 1 and n, and pairs $(\mathfrak{P}, S)$, where $\mathfrak{P} = (P_1, \ldots, P_r)$ is a family of nonintersecting lattice paths which lie below the line $x = y$, $P_i : (i-1, -i+1) \to (n+i, n+2-i)$, $i = 1, \ldots, r$, with the property that if P_r starts with h horizontal steps followed by a vertical step then S is a p-subset of $\{1, 2, \ldots, h-1\}$. (By convention, if $h = 1$ the set $\{1, 2, \ldots, h-1\}$ is meant to be the empty set.) Moreover, Δ_5 satisfies*

$$\text{maj}\,\mathfrak{P} + \|S\| = n(\tau) \qquad \text{if } (\mathfrak{P}, S) = \Delta_5(\tau). \qquad \Box \qquad (5.4.13)$$

PROOF. Let τ be a tableau of the considered type.

First we consider the case that τ has exactly $2r$ columns. We apply the algorithm ROW-DELETE to each of the odd rows of τ, beginning with the lowest and working up to the highest. We thus obtain a sequence $n_1, n_2, \ldots, n_p$ of integers, the n_1 being deleted by the first application of ROW-DELETE, etc. Besides we are left with a tableaux τ_e with only even rows. Clearly, the last element of the first row of τ_e is greater than n_i for all $i = 1, \ldots, p$. Denote this last element of τ_e by h. By the last consideration $\{n_1, n_2, \ldots, n_p\}$ is a subset of $\{1, 2 \ldots, h-1\}$. The observation that the integers $n_1, n_2, \ldots, n_p$ are in increasing order shows that by successively inserting $n_p, n_{p-1}, \ldots, n_1$ into τ_e utilizing ROW-INSERT, the above procedure could be inverted. Hence it is a bijection between tableaux with p odd rows, with exactly $2r$ columns, and with parts between 1 and n, and pairs (τ_e, S), where τ_e is a tableau with even rows, with exactly $2r$ columns, and with parts between 1 and n, with the property that if the last element in the first row of τ_e is h, then S is a p-subset of $\{1, 2, \ldots, h-1\}$. By Proposition 30 these tableaux are in one-to-one correspondence with families $\mathfrak{P} = (P_1, \ldots, P_r)$ of nonintersecting lattice paths which lie below $x = y$, $P_i : (i-1, -i+1) \to (n+i, n+2-i)$, where the path P_r starts with h horizontal steps followed by a vertical one. Note that in the above considerations h, since being an element of τ_e, can at most be n. What is missing is the case $h = n+1$.

Now let the length of the first row of τ be smaller than $2r$. We apply the same method. Use ROW-DELETE repeatedly to get rid of the odd rows of τ. We thus obtain a pair (τ_e, S) where τ_e is a tableau with properties as before but with less than $2r$ columns, and where S is a p-subset of $\{1, 2, \ldots, n\}$. But tableaux τ_e with less than $2r$ columns by Proposition 30 uniquely correspond to families $\mathfrak{P} = (P_1, \ldots, P_r)$ of nonintersecting lattice paths with the properties as before, but where P_r starts with $n+1$ horizontal steps followed by $n+1$ vertical steps. This supplies us with the case $h = n+1$.

Combining both cases we have a bijection between tableaux τ and pairs $(\mathfrak{P}, S)$ of families $\mathfrak{P}$ of nonintersecting lattice paths and p-subsets S as desired. In addition, from (5.4.7) and the fact that S consists of some elements of τ, we derive (5.4.13). $\Box$

Again the same can be done for tableaux with odd parts.

Proposition 33. *There exists a bijection Δ_6 between tableaux τ with p odd rows, with at most $2r$ columns, and with only odd parts which lie between 1 and $2n-1$, and pairs $(\mathfrak{P}, S)$, where $\mathfrak{P} = (P_1, \ldots, P_r)$ is a family of nonintersecting lattice paths which lie below the line $x = y$, $P_i : (2i-2, -2i+2) \to (2n+2i-2, 2n-2i+2)$, $i = 1, \ldots, r$, where the North-East corners of the paths have odd coordinates, with the property that if P_r starts with $2h+1$ horizontal steps followed by a vertical step then S is a p-subset of $\{1, 3, \ldots, 2h-1\}$. (By convention, if $h = 0$ the set $\{1, 3, \ldots, 2h-1\}$ is meant to be the empty set.) Moreover, Δ_6 satisfies*

$$\operatorname{maj} \mathfrak{P} + \|S\| = n(\tau) \qquad \text{if } (\mathfrak{P}, S) = \Delta_6(\tau). \tag{5.4.14}$$

5.5. A determinant lemma. Let the *degree* of a *Laurent polynomial* $p(X) = \sum_{i=M}^{N} a_i x^i$, $M, N \in \mathbb{Z}$, $a_i \in \mathbb{R}$ and $a_N \neq 0$, be defined by $\deg p := N$.

Lemma 34. *Let $X_1, X_2, \ldots, X_r, A_2, A_3, \ldots, A_r, C$ be indeterminates. If $p_0, p_1, \ldots, p_{r-1}$ are Laurent polynomials with $\deg p_j \leq j$ and $p_j(C/X) = p_j(X)$ for $j = 0, 1, \ldots, r-1$, then*

$$\det_{1 \leq s, t \leq r} \left((A_r + X_t) \cdots (A_{s+1} + X_t)(A_r + C/X_t) \cdots (A_{s+1} + C/X_t) \cdot p_{s-1}(X_t) \right)$$

$$= \prod_{1 \leq i < j \leq r} (X_i - X_j)(1 - C/X_i X_j) \prod_{i=1}^{r} A_i^{i-1} \prod_{i=1}^{r} p_{i-1}(-A_i), \tag{5.5.1}$$

with the convention that empty products (like $(A_r + X_t) \cdots (A_{s+1} + X_t)$ for $s = r$) are equal to 1. (The indeterminate A_1, which occurs at the right-hand side of (5.5.1), in fact is superfluous since it occurs in the argument of a constant polynomial.)

PROOF. First we reduce this determinant to a simpler one. Because of $\deg p_{s-1} \leq s-1$ and $p_{s-1}(C/X) = p_{s-1}(X)$ there must exist $\lambda_i^{(s)} \in \mathbb{R}$, $i = 1, \ldots, s-1$, such that

$$p_{s-1}(X) + \sum_{i=1}^{s-1} \lambda_i^{(s)}(A_s + X) \cdots (A_{i+1} + X)(A_s + C/X) \cdots (A_{i+1} + C/X) = \text{constant.}$$

The value of this constant must be $p_{s-1}(-A_s)$. This is seen setting $X = -A_s$.

Hence we obtain that there are $\lambda_i^{(s)} \in \mathbb{R}$ such that

$$p_{s-1}(X) + \sum_{i=1}^{s-1} \lambda_i^{(s)}(A_s + X) \cdots (A_{i+1} + X)(A_s + C/X) \cdots (A_{i+1} + C/X) = p_{s-1}(-A_s). \tag{5.5.2}$$

Now for $j = 2, 3, \ldots, r$ (in that order) we perform the following row operations: Add $\lambda_i^{(j)}/p_{i-1}(-A_i)$ times the i-th row, $i = 1, 2, \ldots, j-1$, to the j-th row. Then by using (5.5.2) it is seen that the t-th entry in the j-th row becomes

$$(A_r + X_t) \cdots (A_{j+1} + X_t)(A_r + C/X_t) \cdots (A_{j+1} + C/X_t) \cdot p_{j-1}(-A_j).$$

This allows us to take $p_{j-1}(-A_j)$ out of the j-th row.

After having completed these operations we obtain that the determinant in (5.5.1) is equal to

$$\prod_{i=1}^{r} p_{i-1}(-A_i) \det_{1 \leq s,t \leq r} \left((A_r + X_t) \cdots (A_{s+1} + X_t)(A_r + C/X_t) \cdots (A_{s+1} + C/X_t) \right).$$

Therefore it remains to prove

$$\det_{1 \leq s,t \leq r} \left((A_r + X_t) \cdots (A_{s+1} + X_t)(A_r + C/X_t) \cdots (A_{s+1} + C/X_t) \right)$$

$$= \prod_{1 \leq i < j \leq r} (X_i - X_j)(1 - C/X_i X_j) \prod_{i=1}^{r} A_i^{i-1} \quad (5.5.3)$$

in order to establish (5.5.1). The determinant on the left-hand side of (5.5.3) is a polynomial in $X_1, \ldots, X_r$ of degree $3\binom{r}{2}$ divided by $\prod_i X_i^{r-1}$. Since, for $1 \leq t_1 < t_2 \leq r$, the determinant is zero for $X_{t_1} = X_{t_2}$ or $X_{t_1} = C/X_{t_2}$, the left-hand side is equal to

$$p(X_1, \ldots, X_r) \prod_{i<j} (X_i - X_j)(X_i X_j - C) \Big/ \prod_i X_i^{r-1},$$

where $p(X_1, \ldots, X_r)$ is some polynomial in the X_i's. Comparing the degrees yields that p is a constant, which is computed by comparing coefficients of $\prod_i X_i^{r-i}$ on both sides of (5.5.3). $\square$

If in the previous lemma we let $C \to 0$, we get the following corollary.

Lemma 35. *Let* $X_1, X_2, \ldots, X_r, A_2, A_3, \ldots, A_r$ *be indeterminates. If* $p_0, p_1,$ $\ldots, p_{r-1}$ *are polynomials with* $\deg p_j \leq j$ *for* $j = 0, 1, \ldots, r-1$, *then*

$$\det_{1 \leq s,t \leq r} \left((A_r + X_t) \cdots (A_{s+1} + X_t) \cdot p_{s-1}(X_t) \right)$$

$$= \prod_{1 \leq i < j \leq r} (X_i - X_j) \prod_{i=1}^{r} p_{i-1}(-A_i). \quad (5.5.4)$$

REMARKS. (1) Note that in Lemma 35 we definitely mean *polynomials* and not *Laurent* polynomials.

(2) In [25, 26, 28] the author has used some determinant lemmas to obtain closed forms for generating functions for several families of plane partitions. In fact, all of these determinant lemmas are special cases of Lemma 34. In particular, Lemma 2.2 of [26] comes out of Lemma 35 by setting

$$p_{j-1}(X) = \prod_{i=1}^{j-1} (B_{i+1} + X).$$

Lemma 7 of [28] is the special case $p_j \equiv 1$, $j = 0, 1, \ldots, r-1$, of Lemma 34.

5.6. A multiple basic hypergeometric summation.

Proposition 36. *For $r \geq 1$ there holds the summation formula*

$$
\sum_{k_1,\ldots,k_r \geq 0} \prod_{i=1}^{r} \left(\frac{\sqrt{q}}{q^i A} \right)^{k_i} \prod_{i=1}^{r} \frac{(m_i A)_{k_i}}{(qm_i/A)_{k_i}}
$$

$$
\times \prod_{1 \leq i < j \leq r} \frac{1 - \frac{m_j}{m_i} q^{k_j - k_i}}{1 - \frac{m_j}{m_i}} \prod_{1 \leq i \leq j \leq r} \frac{1 - m_i m_j q^{k_i + k_j}}{1 - m_i m_j}
$$

$$
= \prod_{1 \leq i < j \leq r} \frac{1 - m_i m_j / q}{1 - m_i m_j} \prod_{i=1}^{r} \frac{(1 - m_i/\sqrt{q})(1 - m_i/A)}{(1 - m_i^2)(1 - \sqrt{q}/q^i A)} , \quad (5.6.1)
$$

provided that there exist nonnegative integers n_i with $n_1 > n_2 > \cdots > n_r$ such that $m_i A = q^{-n_i}$ for all $i = 1, 2, \ldots, r$.

REMARK. I decided to formulate the Proposition in terms of the m_i's rather than the n_i's in order to keep the notation as short as possible. However, in the following the reader should never forget that the m_i's in fact disguise the n_i's via $m_i = q^{-n_i}/A$.

PROOF. It seems to be convenient to first give a rough sketch of what we are going to do. We establish (5.6.1) using a double induction. The main induction is on r. For $r = 1$ equation (5.6.1) can be written in basic hypergeometric notation (cf. subsection 2.1):

$$
{}_6\phi_5 \left[\begin{matrix} m_1^2, qm_1, -qm_1, q, \sqrt{q}m_1, m_1 A \\ m_1, -m_1, m_1^2, \sqrt{q}m_1, qm_1/A \end{matrix} ; q, \frac{1}{\sqrt{q}A} \right] = \frac{(1 - m_1/\sqrt{q})(1 - m_1/A)}{(1 - m_1^2)(1 - 1/\sqrt{q}A)} . \quad (5.6.2)
$$

This is just a special case of the very well-poised ${}_6\phi_5$-summation (3.3.9). (Recall that $m_1 A = q^{-n_1}$ by assumption.) We remark that actually we could start our induction with $r = 0$, thereby avoiding the use of the very well-poised ${}_6\phi_5$-summation. However, it is useful to know what (5.6.1) looks like for $r = 1$, the simplest non-trivial value for r.

Now suppose that (5.6.1) is valid for $r - 1$. To do the induction step, i.e. to settle (5.6.1) for r, provided that it is valid for $r - 1$, we use induction on $n_1 + n_2 + \cdots + n_r$. First we establish a recurrence (identity (5.6.6)) for the left-hand side of (5.6.1). In the second step we shall show that the right-hand side obeys the same recurrence. Hence everything will be done if the validity of (5.6.1) has been checked for the initial values $(n_1, n_2, \ldots, n_{r-1}, 0)$, $n_1 > n_2 > \cdots > n_{r-1} > 0$. Only for this last step it will be necessary to use the induction hypothesis with respect to r.

Let us introduce the following notations: Write $S_r(A; m_1, m_2, \ldots, m_r) = S_r(A; \mathbf{m})$ for the left-hand side and $P_r(A; m_1, m_2, \ldots, m_r) = P_r(A; \mathbf{m})$ for the right-hand side of (5.6.1). Moreover we denote the summand in $S_r(A; \mathbf{m})$ by $E_r(A; m_1, \ldots, m_r, k_1, \ldots, k_r) = E_r(A; \mathbf{m}, \mathbf{k})$ so that $S_r(A; \mathbf{m}) = \sum_{\mathbf{k} \geq \mathbf{0}} E_r(A; \mathbf{m}, \mathbf{k})$. The set of summation indices, $\{k_1, k_2, \ldots, k_r\}$ is denoted by Ω. For a subset T of $\{1, 2, \ldots, r\}$ let Ω_T be the set $\{k_i : i \in T\}$ of corresponding summation indices. For any subset B of Ω the symbol $\sum_{B\geq 1} E_r(A; \mathbf{m}, \mathbf{k})$ means $\sum' E_r(A; \mathbf{m}, \mathbf{k})$, where the sum $\sum'$ is taken over $k_i \geq 1$ for all $i \in B$ and over $k_i \geq 0$ for the remaining $k_i's$.

Since the sum in (5.6.1) by assumption is finite, i.e. consists only of a finite number of terms, the inclusion-exclusion principle can be applied to obtain that with our notations

$$1 = \sum_{T \subseteq \{1,\dots,r\}} (-1)^{|T|} \sum_{\Omega_T \geq 1} E_r(A; \mathbf{m}, \mathbf{k}) . \tag{5.6.3}$$

The 1 on the left-hand side is the term $E_r(A; \mathbf{m}, \mathbf{0})$.

Next for a given $T \subseteq \{1,\dots,r\}$, we express $\sum_{\Omega_T \geq 1} E_r(A; \mathbf{m}, \mathbf{k})$ in terms of $S_r(A; \mathbf{m})$. This is done using the q-augmentation operators $\varepsilon_1, \dots, \varepsilon_r$ which are defined by

$$(\varepsilon_i f)(m_1, \dots, m_i, \dots, m_r) = f(m_1, \dots, q m_i, \dots, m_r)$$

for rational functions $f(m_1, \dots, m_r)$ and $i = 1, \dots, r$. If $n_r \geq 1$ and $q m_t \neq m_{t+1}$ for all t, we get that

$$\sum_{\Omega_T \geq 1} E_r(A; \mathbf{m}, \mathbf{k}) \tag{5.6.4a}$$

$$= \left(\prod_{l \in T} \frac{\sqrt{q}}{q^l A} \frac{(1 - m_l A)}{(1 - q m_l/A)} \right) \frac{\left[\left(\prod_{l \in T} \varepsilon_l \right) \left(\prod_{1 \leq i < j \leq r} (1 - m_j/m_i) \prod_{1 \leq i \leq j \leq r} (1 - m_i m_j) \right) \right]}{\prod_{1 \leq i < j \leq r} (1 - m_j/m_i) \prod_{1 \leq i \leq j \leq r} (1 - m_i m_j)} \tag{5.6.4b}$$

$$\times \left(\prod_{l \in T} \varepsilon_l \right) \sum_{k_1, \dots, k_r \geq 0} E_r(A; \mathbf{m}, \mathbf{k}) . \tag{5.6.4c}$$

On the other hand, suppose that $n_r \geq 1$ and $q m_t = m_{t+1}$ holds for a fixed t. Then $\varepsilon_t E_r(A; \mathbf{m}, \mathbf{k})$ would not be defined due to the denominator term $1 - m_{t+1}/m_t$ in $E_r(A; \mathbf{m}, \mathbf{k})$. Therefore, at first sight, if $t \in T$ but $(t+1) \notin T$ the line (5.6.4c) would not make sense. But in this case we can show that $\sum_{\Omega_T \geq 1} E_r(A; \mathbf{m}, \mathbf{k}) = 0$. This implies that

$$\sum_{\Omega_T \geq 1} E_r(A; \mathbf{m}, \mathbf{k}) = (1 - m_{t+1}/q m_t) R(A; \mathbf{m}) ,$$

where $R(A; \mathbf{m})$ is some rational function with the property that $\lim_{m_{t+1} \to q m_t} R(A; \mathbf{m})$ exists. In fact, the factor $(1 - m_{t+1}/q m_t) = \left(\prod_{l \in T} \varepsilon_l \right) (1 - m_{t+1}/m_t)$ appears in the numerator of (5.6.4b). So the limit $m_{t+1} \to q m_t$ of the rest of (5.6.4b,c) exists, hence also $\lim_{m_{t+1} \to q m_t} \left(\prod_{l \in T} \varepsilon_l \right) \sum_{\Omega_T \geq 1} E_r(A; \mathbf{m}, \mathbf{k})$ exists. This shows that (5.6.4) also holds for $q m_t = m_{t+1}$, $t \in T$, $(t+1) \notin T$, since both sides of (5.6.4) equal zero in this case. To confirm that for $q m_t = m_{t+1}$, $t \in T$, $(t+1) \notin T$ the sum (5.6.4a) indeed vanishes, we write

$$\sum_{\Omega_T \geq 1} E_r(A; \mathbf{m}, \mathbf{k}) = \frac{\sqrt{q}}{q^t A} \frac{(1 - m_t A)}{(1 - q m_t / A)} \tag{5.6.5a}$$

$$\times \sum_{\Omega_{T'} \geq 1} \frac{\left[\varepsilon_t \prod_{i=1}^{r} \left(\frac{\sqrt{q}}{q^j A} \right)^{k_i} \frac{(m_i A)_{k_i}}{(q m_i / A)_{k_i}} \prod_{1 \leq i < j \leq r} (1 - \frac{m_j}{m_i} q^{k_i - k_j}) \prod_{1 \leq i \leq j \leq r} (1 - m_i m_j q^{k_i + k_j}) \right]}{\prod_{1 \leq i < j \leq r} (1 - m_j / m_i) \prod_{1 \leq i \leq j \leq r} (1 - m_i m_j)} \tag{5.6.5b}$$

where $T' = T \backslash \{t\}$. To abbreviate the clumsy expressions, for the moment denote the summand in (5.6.5b) by $F(k_t, k_{t+1})$. The dependence on the other parameters does not concern us for the following considerations. If $k_t = k_{t+1}$ the expression $F(k_t, k_{t+1}) = F(k_t, k_t)$ vanishes because of the numerator term $(1 - \frac{m_{t+1}}{q m_t} q^{k_{t+1} - k_t})$. If $k_{t+1} \neq k_t$ the identity

$$\left(\frac{\sqrt{q}}{q^t A} \right)^{k_t} \left(\frac{\sqrt{q}}{q^{t+1} A} \right)^{k_{t+1}} \frac{1 - \frac{m_{t+1}}{q m_t} q^{k_{t+1} - k_t}}{1 - \frac{m_{t+1}}{m_t}} = - \left(\frac{\sqrt{q}}{q^t A} \right)^{k_{t+1}} \left(\frac{\sqrt{q}}{q^{t+1} A} \right)^{k_t} \frac{1 - \frac{m_{t+1}}{q m_t} q^{k_t - k_{t+1}}}{1 - \frac{m_{t+1}}{m_t}}$$

shows that $F(k_t, k_{t+1}) = -F(k_{t+1}, k_t)$. (Recall that at the moment we are assuming $q m_t = m_{t+1}$. Hence we have $m_{t+1} / q m_t = 1$.) Therefore $\sum_{\Omega_T \geq 1} E_r(A; \mathbf{m}, \mathbf{k})$ is zero whenever there is a t with $q m_t = m_{t+1}$, $t \in T$, and $(t + 1) \notin T$.

Plugging (5.6.4) into (5.6.3) yields the identity

$$1 = \sum_{T \subseteq \{1, \ldots, r\}} (-1)^{|T|} L_T(A; \mathbf{m}) \cdot \left(\prod_{l \in T} \varepsilon_l \right) S_r(A; \mathbf{m}), \tag{5.6.6}$$

where $L_T(A; \mathbf{m})$ is the expression in (5.6.4b). The previous considerations guarantee that (5.6.6) holds for all $A, m_1, \ldots, m_r$ such that $n_1 > n_2 > \cdots > n_r \geq 1$.

Next it is shown that the right-hand side $P(A; \mathbf{m})$ of (5.6.1) also satisfies (5.6.6). This is done by a verification. In (5.6.6) replace $S(A; \mathbf{m})$ by $P(A; \mathbf{m})$. After some manipulation this leads to

$$1 = \left(\prod_{1 \leq i < j \leq r} \frac{1 - m_i m_j / q}{1 - m_i m_j} \right) \left(\prod_{i=1}^{r} \frac{(1 - m_i / \sqrt{q})(1 - m_i / A)}{(1 - m_i^2)(1 - \sqrt{q} / q^i A)} \right)$$

$$\times \sum_{T \subseteq \{1, \ldots, r\}} (-1)^{|T|} \prod_{j \in T} \left(\frac{\sqrt{q}}{q^j A} \frac{1 - m_j A}{1 - m_j / A} \right)$$

$$\times \frac{\left[\left(\prod_{l \in T} \varepsilon_l \right) \left(\prod_{1 \leq i < j \leq r} (1 - m_j / m_i)(1 - m_i m_j / q) \prod_{i=1}^{r} (1 - m_i / \sqrt{q}) \right) \right]}{\prod_{1 \leq i < j \leq r} (1 - m_j / m_i)(1 - m_i m_j / q) \prod_{i=1}^{r} (1 - m_i / \sqrt{q})}. \tag{5.6.7}$$

Multiplication of both sides of (5.6.7) by the denominator of the right-hand side gives

$$
\prod_{1 \le i < j \le r} \left((1 - m_i m_j)(m_i - m_j) \right) \prod_{i=1}^{r} \left((1 - m_i^2)(1 - \sqrt{q}/q^i A) \right)
$$

$$
= \sum_{T \subseteq \{1,\ldots,r\}} (-1)^{|T|} \prod_{j \in T} \left(\frac{\sqrt{q}}{q^j A} \frac{(1 - m_j A)}{(1 - m_j/A)} \right) \prod_{i=1}^{r} \left(m_i^{r-i}(1 - m_i/A) \right)
$$

$$
\times \left[\left(\prod_{l \in T} \varepsilon_l \right) \left(\prod_{1 \le i < j \le r} (1 - m_j/m_i)(1 - m_i m_j/q) \prod_{i=1}^{r} (1 - m_i/\sqrt{q}) \right) \right]
$$

$$
=: \sum_{T \subseteq \{1,\ldots,r\}} \mathrm{Expr}(T) \,. \tag{5.6.8}
$$

The right-hand side of (5.6.8) is a polynomial in $m_1, m_2, \ldots, m_r$ of total degree $2r + 3\binom{r}{2}$. We are going to show that the right-hand side vanishes for $m_s = m_t$, for $m_s = 1/m_t$, and for $m_t = \pm 1$. Suppose that this is done. It follows that

$$
\prod_{1 \le i < j \le r} \left((1 - m_i m_j)(m_i - m_j) \right) \prod_{i=1}^{r} (1 - m_i^2) \tag{5.6.9}
$$

must be a factor of the right-hand side. This factor also has total degree $2r + 3\binom{r}{2}$ in $m_1, m_2, \ldots, m_r$. Therefore the right-hand side of (5.6.8) is the expression in (5.6.9) times some expression which is independent of $m_1, \ldots, m_r$. To get this expression, look at the coefficient of $\prod_{i=1}^{r} m_i^{r-i}$ in the right-hand side of (5.6.8). It is

$$
\sum_{T \subseteq \{1,\ldots,r\}} (-1)^{|T|} \prod_{j \in T} \frac{\sqrt{q}}{q^j A} = \prod_{i=1}^{r} (1 - \sqrt{q}/q^i A) \,,
$$

just as desired.

Let $m_t = \pm 1$ for a fixed t. Moreover, fix a subset T_1 of $\{1, 2, \ldots, r\}$ with $t \notin T_1$ and write $T_2 := T_1 \cup \{t\}$. Then it is seen that

$$
\mathrm{Expr}(T_1)\big|_{m_t=1} = - \mathrm{Expr}(T_2)\big|_{m_t=1} \,,
$$

and the same holding if 1 is replaced by -1. ($\mathrm{Expr}(T)$ was defined in (5.6.8) as the summand of the sum on the right-hand side. $\mathrm{Expr}(T)\big|_{m_t=1}$ means the evaluation of

Expr(T) at $m_t = 1$.) Namely for $m_t = 1$,

$$\mathrm{Expr}(T_2)\big|_{m_t=1} = -(-1)^{|T_1|}\frac{\sqrt{q}}{q^t A}\Big(\prod_{j\in T_1}\frac{\sqrt{q}}{q^j A}\frac{(1-m_j A)}{(1-m_j/A)}\Big)\cdot\frac{1-A}{1-1/A}$$

$$\times\Big(\prod_{i=1}^{r}m_i^{r-i}(1-m_i/A)\Big)\Big|_{m_t=1}\times\Big[\big(\prod_{l\in T_1}\varepsilon_l\big)\big(\prod_{i=1}^{r}(1-m_i/\sqrt{q})\big)$$

$$\prod_{1\le i<j\le r}(1-m_i/m_j)(1-m_im_j/q)$$

$$\frac{1-\sqrt{q}}{1-1/\sqrt{q}}\prod_{i=1}^{t-1}\frac{1-q/m_i}{1-1/m_i}\prod_{i=t+1}^{r}\frac{1-m_i/q}{1-m_i}\prod_{\substack{i=1\\i\ne t}}^{r}\frac{1-m_i}{1-m_i/q}\Big]\Big|_{m_t=1}$$

$$= -\mathrm{Expr}(T_1)\big|_{m_t=1}\ .$$

A similar computation can be done for $m_t = -1$. Hence for $t\notin T_1$ each two terms $\mathrm{Expr}(T_1)\big|_{m_t=1}$ and $\mathrm{Expr}(T_1\cup\{t\})\big|_{m_t=1}$ cancel in the sum on the right-hand side of (5.6.8) when it is evaluated at $m_t = 1$, the same being true for $m_t = -1$. Therefore it must be zero for $m_t = \pm 1$.

Now we turn to the case $m_s = m_t$ for fixed s and t, $1\le s < t\le r$. This obviously only makes sense for $r\ge 2$. Consider a set $S_1\subseteq\{1,\dots,r\}$ with $s,t\notin S_1$. Let us set $S_2 := S_1\cup\{s\}$, $S_3 := S_1\cup\{t\}$, and $S_4 := S_1\cup\{s,t\}$. We have

$$\mathrm{Expr}(S_2)\big|_{m_s=m_t} = (-1)^{|S_3|}q^{t-s}\prod_{j\in S_3}\Big(\frac{\sqrt{q}}{q^j A}\frac{(1-m_j A)}{(1-m_j/A)}\Big)\prod_{i=1}^{r}\big(m_i^{r-i}(1-m_i/A)\big)\big|_{m_s=m_t}$$

$$(5.6.10\mathrm{a})$$

$$\times\Big[\big(\prod_{l\in S_1}\varepsilon_l\big)\prod_{1\le i<j\le r}\big((1-m_j/m_i)(1-m_im_j/q)\big)\prod_{i=1}^{r}(1-m_i/\sqrt{q})$$

$$(5.6.10\mathrm{b})$$

$$\Big(\prod_{i=1}^{s-1}\frac{1-qm_s/m_i}{1-m_s/m_i}\prod_{\substack{i=s+1\\i\ne t}}^{r}\frac{1-m_i/qm_s}{1-m_i/m_s}\Big)\frac{1-m_t/qm_s}{1-m_t/m_s}$$

$$(5.6.10\mathrm{c})$$

$$\Big(\prod_{i=1}^{r}\frac{1-m_im_s}{1-m_im_s/q}\Big)\frac{1-m_s\sqrt{q}}{1-m_s/\sqrt{q}}\Big]\Big|_{m_s=m_t}\ .$$

$$(5.6.10\mathrm{d})$$

(The term $(1 - m_t/m_s)$ in the denominator of (5.6.10c) seems to be a problem. But it is not since in fact it cancels out with a term of (5.6.10b).) The expression $\prod_{\substack{i=s+1\\i\ne t}}^{r}\frac{1-m_i/qm_s}{1-m_i/m_s}$ has to be split up into

$$\prod_{i=s+1}^{t-1}\frac{1-m_i/qm_s}{1-m_i/m_s}\prod_{i=t+1}^{r}\frac{1-m_i/qm_s}{1-m_i/m_s}\ .$$

A little bit of manipulation of the first product turns this into

$$q^{s-t+1} \prod_{i=s+1}^{t-1} \frac{1-qm_s/m_i}{1-m_s/m_i} \prod_{i=t+1}^{r} \frac{1-m_i/qm_s}{1-m_i/m_s} \ .$$

Moreover, we have

$$(1-m_t/qm_s)\big|_{m_s=m_t} = -\frac{1}{q}(1-qm_t/m_s)\big|_{m_s=m_t} \ .$$

Taking these manipulations into account, we finally obtain

$$\mathrm{Expr}(S_2)\big|_{m_s=m_t} = q^{t-s} \cdot q^{s-t+1} \cdot \left(-\frac{1}{q}\right) \cdot \mathrm{Expr}(S_3)$$

$$= -\,\mathrm{Expr}(S_3) \ .$$

Since obviously $\mathrm{Expr}(S_1)\big|_{m_s=m_t} = \mathrm{Expr}(S_4)\big|_{m_s=m_t} = 0$, this means that for $s,t \notin S_1$ each four terms $\mathrm{Expr}(S_1)\big|_{m_s=m_t}$, $\mathrm{Expr}(S_1 \cup \{s\})\big|_{m_s=m_t}$, $\mathrm{Expr}(S_1 \cup \{t\})\big|_{m_s=m_t}$, $\mathrm{Expr}(S_1 \cup \{s,t\})\big|_{m_s=m_t}$ cancel in the sum on the right-hand side of (5.6.8) when it is evaluated at $m_s = m_t$. Therefore it must be zero for $m_s = m_t$.

The arguments for $m_s = 1/m_t$ are similar. This time one obtains

$$\mathrm{Expr}(S_2)\big|_{m_s=1/m_t} = \mathrm{Expr}(S_3)\big|_{m_s=1/m_t} = 0$$

and

$$\mathrm{Expr}(S_1)\big|_{m_s=1/m_t} = -\,\mathrm{Expr}(S_4)\big|_{m_s=1/m_t}.$$

Summarizing, we have shown that both sides of (5.6.1) satisfy the functional relation (5.6.6). To be precise, for all $n_1 > n_2 > \cdots > n_r \geq 1$ we proved that

$$1 = \sum_{T \subseteq \{1,\dots,r\}} (-1)^{|T|} L_T(A;q^{-\mathbf{n}}/A) \cdot \left(\prod_{l \in T} E_l^{-1}\right) S_r(A;q^{-\mathbf{n}}/A) \qquad (5.6.11a)$$

and

$$1 = \sum_{T \subseteq \{1,\dots,r\}} (-1)^{|T|} L_T(A;q^{-\mathbf{n}}/A) \cdot \left(\prod_{l \in T} E_l^{-1}\right) P_r(A;q^{-\mathbf{n}}/A) \ , \qquad (5.6.11b)$$

where $q^{-\mathbf{n}}/A$ is a short-hand notation for $q^{-n_1}/A, q^{-n_2}/A, \dots, q^{-n_r}/A$. E_l is the usual shift operator with respect to n_l,

$$(E_l f)(m_1,\dots,m_l,\dots,m_r) = f(m_1,\dots,m_l+1,\dots,m_r) \ .$$

Besides, the expressions of the form $S_r(A;q^{-\hat{\mathbf{n}}}/A)$ and $P_r(A;q^{-\hat{\mathbf{n}}}/A)$ which are involved in the non-zero terms of (5.6.11) all satisfy $\hat{n}_1 > \hat{n}_2 > \cdots > \hat{n}_r \geq 0$. Therefore induction on $n_1 + n_2 + \cdots + n_r$ can be performed, as soon as the validity of (5.6.1) for the initial values $(n_1, n_2, \ldots, n_{r-1}, 0)$, $n_1 > \cdots > n_{r-1} > 0$ is checked.

Let $n_r = 0$, which is equivalent to $m_r = 1/A$. It is clear that in this case there are nonvanishing terms in the sum at the left-hand side of (5.6.1) only if $k_r = 0$. Hence,

$$S_r(A; m_1, \ldots, m_{r-1}, 1/A)$$

$$= \sum_{k_1, \ldots, k_{r-1} \geq 0} \prod_{i=1}^{r-1} \left(\frac{\sqrt{q}}{q^i A}\right)^{k_i} \prod_{i=1}^{r-1} \frac{(m_i A)_{k_i}}{(q m_i / A)_{k_i}}$$

$$\times \prod_{1 \leq i < j \leq r-1} \frac{1 - \frac{m_j}{m_i} q^{k_j - k_i}}{1 - \frac{m_j}{m_i}} \prod_{1 \leq i \leq j \leq r-1} \frac{1 - m_i m_j q^{k_i + k_j}}{1 - m_i m_j}$$

$$\times \prod_{i=1}^{r-1} \frac{1 - q^{-k_i}/m_i A}{1 - 1/m_i A} \frac{1 - m_i q^{k_i}/A}{1 - m_i/A}$$

$$= S_{r-1}(qA; m_1, \ldots, m_{r-1})$$

$$= P_{r-1}(qA; m_1, \ldots, m_{r-1})$$

$$= P_r(A; m_1, \ldots, m_{r-1}, 1/A) .$$

The next to the last step used the induction hypothesis with respect to r, the last step is an easy exercise.

The proof of the Proposition now is complete. $\square$

References

1. G. E. Andrews, *The Theory of Partitions*, Encyclopedia of Mathematics and its Applications, Vol. 2, Addison–Wesley, Reading, 1976.

2. G. E. Andrews, *Plane partitions I: The MacMahon conjecture*, Studies in foundations and combinatorics, G.-C. Rota ed., Adv. in Math. Suppl. Studies, Vol. 1, 1978, pp. 131–150.

3. G. E. Andrews, *Plane partitions II: The equivalence of the Bender–Knuth and MacMahon conjectures*, Pacific J. Math. **72** (1977), 283–291.

4. E. A. Bender, D. E. Knuth, *Enumeration of plane partitions*, J. Combin. Theory A **13** (1972), 40–54.

5. N. Bourbaki, *Groupes et algèbres de Lie*, Chapitres IV, V, VI, Éléments de Mathématiques, Fasc. 34, Hermann, Paris, 1968.

6. W. H. Burge, *Four correspondences between graphs and generalized Young tableaux*, J. Combin. Theory A **17** (1974), 12–30.

7. S. H. Choi, D. Gouyou–Beauchamps, *Enumération de tableaux de Young semi-standard*, Actes de 3eme Colloque de Series Formelles et Combinatoire Algebrique, Bordeaux 1991, M. Delest, G. Jacob, P. Leroux, Eds., LabRI, Université Bordeaux I, 1991, pp. 229–243.

8. L. Comtet, *Advanced Combinatorics*, D. Reidel, Dordrecht, Holland, 1974.

9. R. Y. Denis, R. A. Gustafson, *An $SU(n)$ q-beta integral transformation and multiple hypergeometric series identities*, SIAM J. Math. Anal. **23** (1992), 552–561.

10. M. Desainte–Catherine, G. Viennot, *Enumeration of certain Young tableaux with bounded height*, Combinatoire énumérative, G. Labelle, P. Leroux, Eds., Springer–Verlag, Berlin, Heidelberg, New York, 1986, pp. 58–67.

11. J. Désarménien, *La démonstration des identités de Gordon et MacMahon et de deux identités nouvelles*, Actes de 15e Séminaire Lotharingien, I. R. M. A. Strasbourg, 1987, pp. 39–49.

12. J. Désarménien, *Une généralisation des fomules de Gordon et de MacMahon*, Comptes Rendus Acad. Sci. Paris, Série I **309** (1989), 269–272.

13. J. Fürlinger, J. Hofbauer, *q-Catalan numbers*, J. Combin. Theory A **40** (1985), 248–264.

14. G. Gasper, M. Rahman, *Basic hypergeometric series*, Encyclopedia of Mathematics And Its Applications 35, Cambridge University Press, Cambridge, 1990.

15. I. M. Gessel, *Symmetric functions and P-recusiveness*, J. Combin. Theory A **53** (1990), 257–285.

16. I. M. Gessel, G. Viennot, *Binomial determinants, paths, and hook length formulae*, Adv. in Math. **58** (1985), 300–321.

17. I. M. Gessel, G. Viennot, *Determinants, paths, and plane partitions*, preprint.

18. B. Gordon, *A proof of the Bender–Knuth conjecture*, Pacific J. Math. **108** (1983), 99–113.

19. I. P. Goulden, *A linear operator for symmetric functions and tableaux in a strip with given trace*, Discrete Math. **99** (1992), 69–77.

20. S. Guha, S. Padmanabhan, *A new derivation of the generating function for the major index*, Discrete Math. **81** (1990), 211–215.

21. R. A. Gustafson, *The Macdonald identities for affine root systems of classical type and hypergeometric series very-well-poised on semisimple Lie algebras*, Ramanujan International Symposium on Analysis (Dec. 26th to 28th, 1987, Pune, India), N. K. Thakare, ed., 1989, pp. 187–224.

22. R. A. Gustafson, private communication.

23. J. E. Humphreys, *Introduction to Lie algebras and representation theory*, Springer–Verlag, Berlin, Heidelberg, New York, 1972.

24. D. E. Knuth, *Permutations, matrices, and generalized Young tableaux*, Pacific J. Math. **34** (1970), 709–727.

25. C. Krattenthaler, *Enumeration of lattice paths and generating functions for skew plane partitions*, Manuscripta Math. **63** (1989), 129–156.

26. C. Krattenthaler, *Generating functions for plane partitions of a given shape*, Manuscripta Math. **69** (1990), 173–202.

27. C. Krattenthaler, *Counting lattice paths with a linear boundary, Part 2: q-ballot and q-Catalan numbers*, Sitz.ber. d. ÖAW, Math-naturwiss. Klasse **198** (1989), 171–199.

28. C. Krattenthaler, *Generating functions for shifted plane partitions*, J. Statist. Plann. Inference **34** (1993), 197–208.

29. C. Krattenthaler, *Non-crossing two-rowed arrays*, in preparation; a preliminary report appeared in: Proc. of the 5th Conference on Formal Power Series and Algebraic Combinatorics, Florence, 1993 (A. Barlotti, M. Delest, R. Pinzani, eds.), D.S.I., Università di Firenze, pp. 301–314.

30. G. M. Lilly, S. C. Milne, *The C_l Bailey transform and Bailey Lemma*, Constructive Approximation **9** (1993), 473–500.

31. G. M. Lilly, S. C. Milne, *Consequences of the A_l and C_l Bailey Transform and Bailey Lemma*, Proceedings of the 4th Colloque de Series Formelles et Combinatoire Algebrique, Montreal 1992, P. Leroux, C. Reutenauer, eds., LaCIM, Université Québec à Montréal, 1992, pp. 297–312.

32. P. A. MacMahon, *Combinatory Analysis*, Vol. 2, (Cambridge University Press, Cambridge, 1916; reprinted by Chelsea, New York, 1960).

33. I. G. Macdonald, *Symmetric Functions and Hall Polynomials*, Oxford University Press, New York/London, 1979.

34. S. C. Milne, *The multidimensional $_1\psi_1$ sum and Macdonald identities for $A_l^{(1)}$*, Theta Functions Bourdoin 1987, L. Ehrenpreis and R. C. Gunning, eds., Prooceedings of Symposia in Pure Mathematics, Vol. 49, Part 2, 1989, pp. 323–359.

35. S. C. Milne, *Classical partition functions and the $U(n + 1)$ Rogers–Selberg identity*, Discrete Math. **99** (1992), 199–246.

36. R. A. Proctor, *Bruhat lattices, plane partitions generating functions, and minuscule representations*, Europ. J. Combin. **5** (1984), 331–350.

37. R. A. Proctor, *New symmetric plane partition identities from invariant theory work of DeConcini and Procesi*, Europ. J. Combin. **11** (1990), 289–300.

38. B. E. Sagan, *The symmetric group*, Wadsworth & Brooks/Cole, Pacific Grove, California, 1991.

39. R. P. Stanley, *Theory and applications of plane partitions: Part 1,2*, Stud. Appl. Math **50** (1971), 167–188, 259–279.

40. J. Stembridge, *Hall–Littlewood functions, plane partitions and the Rogers–Ramanujan identities*, Trans. Amer. Math. Soc. **319** (1990), 469–498.

41. J. Stembridge, *Nonintersecting lattice paths, pfaffians and plane partitions*, Adv. Math. **83** (1990), 96–131.

42. G. Viennot, *Une forme géométrique de la correspondance de Robinson–Schensted*, Combinatoire et Représentation du Groupe Symétrique, D. Foata ed., Lecture Notes in Math., Vol. 579, Springer–Verlag, New York, 1977, pp. 29–58.

Editorial Information

To be published in the *Memoirs*, a paper must be correct, new, nontrivial, and significant. Further, it must be well written and of interest to a substantial number of mathematicians. Piecemeal results, such as an inconclusive step toward an unproved major theorem or a minor variation on a known result, are in general not acceptable for publication. *Transactions* Editors shall solicit and encourage publication of worthy papers. Papers appearing in *Memoirs* are generally longer than those appearing in *Transactions* with which it shares an editorial committee.

As of February 1, 1995, the backlog for this journal was approximately 5 volumes. This estimate is the result of dividing the number of manuscripts for this journal in the Providence office that have not yet gone to the printer on the above date by the average number of monographs per volume over the previous twelve months, reduced by the number of issues published in four months (the time necessary for preparing an issue for the printer). (There are 6 volumes per year, each containing at least 4 numbers.)

A Copyright Transfer Agreement is required before a paper will be published in this journal. By submitting a paper to this journal, authors certify that the manuscript has not been submitted to nor is it under consideration for publication by another journal, conference proceedings, or similar publication.

Information for Authors and Editors

Memoirs are printed by photo-offset from camera copy fully prepared by the author. This means that the finished book will look exactly like the copy submitted.

The paper must contain a *descriptive title* and an *abstract* that summarizes the article in language suitable for workers in the general field (algebra, analysis, etc.). The *descriptive title* should be short, but informative; useless or vague phrases such as "some remarks about" or "concerning" should be avoided. The *abstract* should be at least one complete sentence, and at most 300 words. Included with the footnotes to the paper, there should be the 1991 *Mathematics Subject Classification* representing the primary and secondary subjects of the article. This may be followed by a list of *key words and phrases* describing the subject matter of the article and taken from it. A list of the numbers may be found in the annual index of *Mathematical Reviews*, published with the December issue starting in 1990, as well as from the electronic service e-MATH [**telnet e-MATH.ams.org** (or **telnet 130.44.1.100**). Login and password are **e-math**]. For journal abbreviations used in bibliographies, see the list of serials in the latest *Mathematical Reviews* annual index. When the manuscript is submitted, authors should supply the editor with electronic addresses if available. These will be printed after the postal address at the end of each article.

Electronically prepared manuscripts. The AMS encourages submission of electronically prepared manuscripts in $\mathcal{A}_{\mathcal{M}}\mathcal{S}$-TEX or $\mathcal{A}_{\mathcal{M}}\mathcal{S}$-LATEX because properly prepared electronic manuscripts save the author proofreading time and move more quickly through the production process. To this end, the Society has prepared "preprint" style files, specifically the amsppt style of $\mathcal{A}_{\mathcal{M}}\mathcal{S}$-TEX and the amsart style of $\mathcal{A}_{\mathcal{M}}\mathcal{S}$-LATEX, which will simplify the work of authors and of the production staff. Those authors who make use of these style files from the beginning of the writing process will further reduce their own effort. Electronically submitted manuscripts prepared in plain TeX or LaTeX do not mesh properly with the AMS production systems and cannot, therefore, realize the same kind of expedited processing. Users of plain TeX should have little difficulty learning $\mathcal{A}_{\mathcal{M}}\mathcal{S}$-TeX, and LaTeX users will find that $\mathcal{A}_{\mathcal{M}}\mathcal{S}$-LaTeX is the same as LaTeX with additional commands to simplify the typesetting of mathematics.

Guidelines for Preparing Electronic Manuscripts provides additional assistance and is available for use with either $\mathcal{A}_{\mathcal{M}}\mathcal{S}$-TeX or $\mathcal{A}_{\mathcal{M}}\mathcal{S}$-LaTeX. Authors with FTP access may obtain *Guidelines* from the Society's Internet node e-MATH.ams.org (130.44.1.100). For those without FTP access *Guidelines* can be obtained free of charge from the e-mail address guide-elec@ math.ams.org (Internet) or from the Customer Services Department, American Mathematical Society, P.O. Box 6248, Providence, RI 02940-6248. When requesting *Guidelines*, please specify which version you want.

At the time of submission, authors should indicate if the paper has been prepared using $\mathcal{A}_{\mathcal{M}}\mathcal{S}$-TeX or $\mathcal{A}_{\mathcal{M}}\mathcal{S}$-LaTeX. The *Manual for Authors of Mathematical Papers* should be consulted for symbols and style conventions. The *Manual* may be obtained free of charge from the e-mail address cust-serv@math.ams.org or from the Customer Services Department, American Mathematical Society, P.O. Box 6248, Providence, RI 02940-6248. The Providence office should be supplied with a manuscript that corresponds to the electronic file being submitted.

Electronic manuscripts should be sent to the Providence office immediately after the paper has been accepted for publication. They can be sent via e-mail to pub-submit@math.ams.org (Internet) or on diskettes to the Publications Department, American Mathematical Society, P.O. Box 6248, Providence, RI 02940-6248. When submitting electronic manuscripts please be sure to include a message indicating in which publication the paper has been accepted.

Two copies of the paper should be sent directly to the appropriate Editor and the author should keep one copy. The *Guide for Authors of Memoirs* gives detailed information on preparing papers for *Memoirs* and may be obtained free of charge from the Editorial Department, American Mathematical Society, P.O. Box 6248, Providence, RI 02940-6248. For papers not prepared electronically, model paper may also be obtained free of charge from the Editorial Department.

Any inquiries concerning a paper that has been accepted for publication should be sent directly to the Editorial Department, American Mathematical Society, P.O. Box 6248, Providence, RI 02940-6248.

Recent Titles in This Series

(*Continued from the front of this publication*)

(See the AMS catalog for earlier titles)